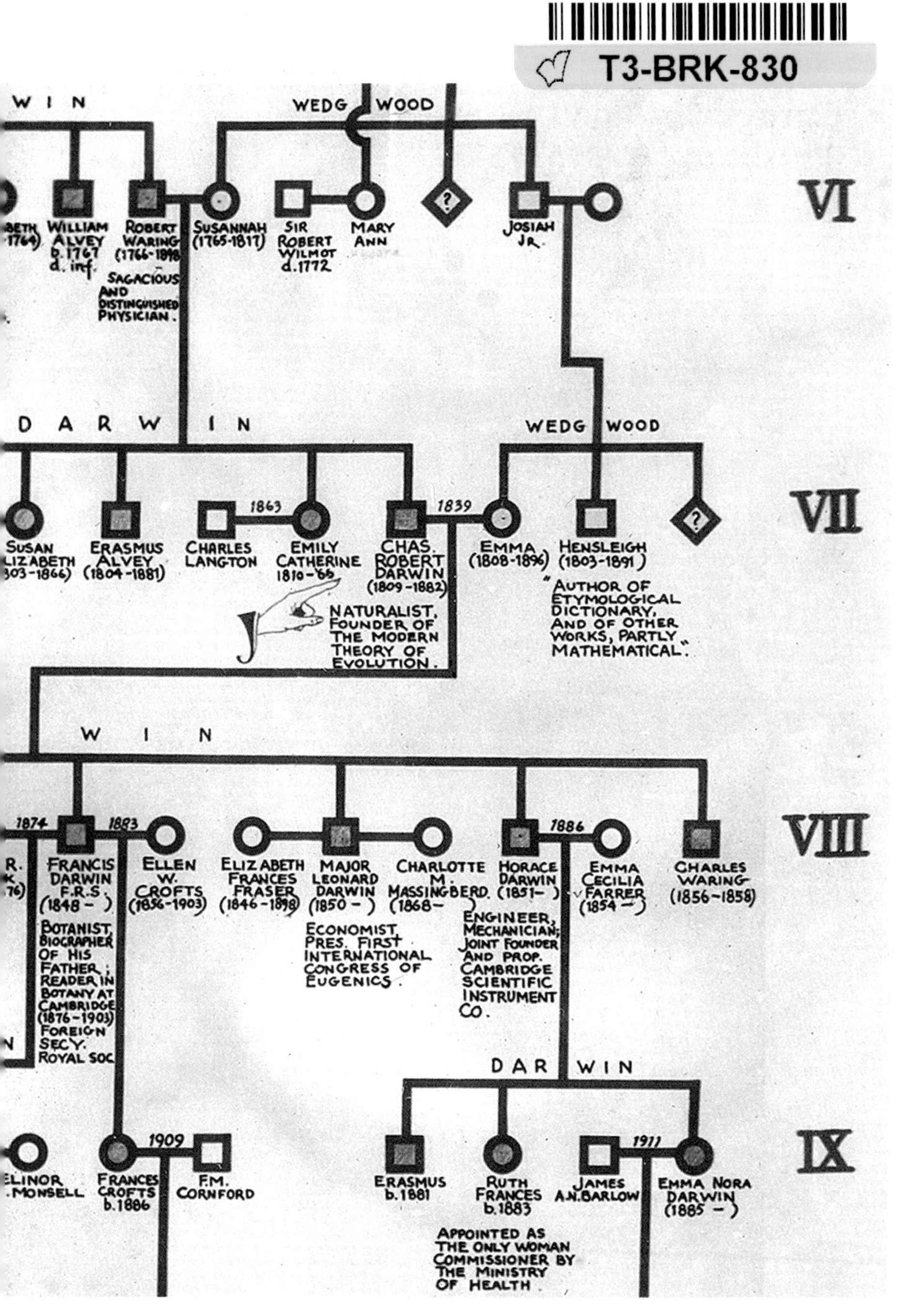

ongress of Eugenics at the American Museum of Natural History in New York City in 1932. It was

DARWIN AND HIS CHILDREN

DARWIN AND HIS CHILDREN

His Other Legacy

TIM M. BERRA

Oxford University Press is a department of the University of Oxford.
It furthers the University's objective of excellence in research, scholarship,
and education by publishing worldwide.

Oxford New York
Auckland Cape Town Dar es Salaam Hong Kong Karachi
Kuala Lumpur Madrid Melbourne Mexico City Nairobi
New Delhi Shanghai Taipei Toronto

With offices in
Argentina Austria Brazil Chile Czech Republic France Greece
Guatemala Hungary Italy Japan Poland Portugal Singapore
South Korea Switzerland Thailand Turkey Ukraine Vietnam

Published in the United States of America by
Oxford University Press
198 Madison Avenue, New York, NY 10016

Library of Congress Cataloging-in-Publication Data
Berra, Tim M., 1943–
Darwin and his children: his other legacy / Tim M. Berra.
pages cm
Includes bibliographical references and index.
ISBN 978–0–19–930944–3 (alk. paper)
1. Darwin, Charles, 1809–1882. 2. Darwin, Charles, 1809–1882–
Family. 3. Darwin family. 4. Naturalists–England–Biography. 5. Evolution
(Biology) I. Title.
QH31.D2B474 2013
576.8'2–dc23
2013003193

9 8 7 6 5 4 3 2 1
Printed in the United States of America
on acid-free paper

Dedicated to:

The Fulbright Program,
The Ohio State University,
The Museums and Art Galleries
of the Northern Territory, and
Charles Darwin University;
educational entities that have made
my career not only possible,
but also a great deal of fun.

CONTENTS

PREFACE

The name of the game in evolution is to get your genes into future generations (Berra 1990). Of course Charles Robert Darwin had no knowledge of modern genetics. Nevertheless, to that end Darwin and his wife, Emma née Wedgwood Darwin, who was his first cousin, had 10 children. Three of these children, Anne Elizabeth, Mary Eleanor, and Charles Waring, died in childhood. One daughter, Elizabeth ("Bessy"), did not marry. The other six Darwin offspring had long-term marriages. Three of these marriages yielded no offspring (William Erasmus, Henrietta Emma née Darwin Litchfield, and Leonard), while George, Francis, and Horace provided grandchildren.

I have been fascinated by the life of Darwin since reading *The Voyage of the* Beagle as a high school biology student, and I became interested in his children as I was doing research for my little volume *Charles Darwin: The Concise Story of an Extraordinary Man*. This led me to investigate the consanguineous marriages within the Darwin-Wedgwood families. My Spanish geneticist colleagues, Gonzalo Alvarez and Francisco C. Ceballos, and I showed that Darwin's children were subject to a moderate level of inbreeding similar to any marriage between first cousins (Berra et al. 2010a). Furthermore, we showed a remarkable inbreeding depression for child survival in the Darwin-Wedgwood dynasty. A recent reanalysis of our data shows an important inbreeding depression for male fertility in a number of Darwin-Wedgwood marriages (Alvarez and Ceballos, personal communication). About 10 percent of the marriages in upper-class English society at that time were similarly consanguineous (Kuper 2009). We now know that susceptibility to bacterial infections, the cause of death for Anne and Charles Waring, (Lyons et al. 2009a, 2009b) and unexplained infertility (Golubovsky 2008) are considered possible consequences of consanguinity in Darwin's children.

On the other hand, it was not all genetic doom and gloom. Three of Darwin's children (the same three who left offspring: George, Francis, and Horace) were elected Fellows of the Royal Society for their scientific accomplishments and were knighted. This latter honor escaped Charles, as he was much too controversial for Queen Victoria's taste, even though her husband, Prince Albert, supported Lord Palmerston's nomination of Charles for the Honors List in June 1859. The Bishop of Oxford, Samuel Wilberforce, effectively blocked the nomination (Desmond and Moore 1991).

Further investigation led to the rediscovery of a huge pedigree of the Galton-Darwin-Wedgwood families, with ancestors dating to the early 1600s, that was originally exhibited at the Third International Congress of Eugenics held at the American Museum of Natural History in 1932 (Berra et al. 2010b). A portion of this pedigree is reproduced on the front endpapers to serve as a scorecard for the players in the lineup of this book. The rear endpaper contains a map of the localities in the United Kingdom that are significant with regard to the various Darwin family members.

My interest in the Darwin children was further piqued when I was the keynote speaker at a Charles Darwin symposium celebrating the two-hundredth anniversary of his birth. This symposium was sponsored by Charles Darwin University and held at the Darwin Convention Centre in Darwin, Northern Territory, Australia, on 22–24 September 2009. Just when I thought it could not get any more Darwin than that, I was informed that Charles Darwin's great-great-grandson, Chris Darwin (great-grandson of George Darwin and grandson of William Darwin) was in the audience.

In addition to my "day job" as professor emeritus of Evolution, Ecology, and Organismal Biology at The Ohio State University, I am proud to say that I hold the titles of University Professorial Fellow at Charles Darwin University and Research Associate at the Museums and Art Galleries of the Northern Territory in Darwin, Australia. Charles Darwin is never far from my thoughts when I am in this most beautiful, tropical namesake city of Darwin.

The first chapter in the present volume is a brief overview of Charles Darwin's life, accomplishments, and the paradigm shift his ideas induced. What then follows is a concise summary of each of Darwin's children's lives, which I have gleaned from published accounts. The life of the Darwin children growing up at Down House is familiar ground that has been plowed by the major Darwin biographers: Bowlby 1990; Desmond and Moore 1991; Browne 1995, 2002; Healey 2001; R. Keynes 2001; Colp 2008; Loy and Loy 2010; and others using the same sources, such as the version of Darwin's *Autobiography* with previous omissions restored (Barlow 1958), *The Correspondence of Charles Darwin* (Burkhardt et al. 1985–), *The Life and Letters of Charles Darwin* (F. Darwin 1887), *Emma Darwin: A Century of Family Letters* (H. Litchfield 1915), *Period Piece* (Raverat 1952), and the ever-useful *Charles Darwin: A Companion* (Freeman 1978). In a commemorative volume published on the one-hundredth anniversary of Darwin's death, Freeman (1982) provided insights into the private lives of the Darwin family. No less a source than "Darwin's Bulldog" himself, Thomas Henry Huxley, has written about the intellectual evolution of his friend and confidant (T. Huxley 1888). I have relied on these references, but I have chosen not to interrupt the flow of the narrative by continually inserting citations to these standard works at every factual statement. Rather, I have sprinkled them around to indicate where the reader can explore the topic further. Other specific references are cited where useful.

I am grateful to Ohio State University librarians Bruce Leach and Amanda Maddock for their help with obtaining references. Kate Shannon arranged the pedigree (front endpapers), Figure 4.2 and the map (rear endpaper), and constructed the timeline (appendix 1). The vast majority of the photographs in this book are in the public domain. I have indicated the owner of the photographs when known.

I thank science editor Jeremy Lewis for giving me the opportunity to be an Oxford University Press author. The manuscript benefited greatly from the expert suggestions of Darwin scholar Duncan M. Porter and several anonymous reviewers. Vince Burke encouraged me to tell the story of Darwin's children. Project manager Abidha Sulaiman kept the production process moving along efficiently. I am especially grateful to copyeditor Kathleen Capels. This is the second book she has fine-tuned for me. Her skill as a wordsmith, her attention to detail, and her biological knowledge and enthusiasm were a perfect combination for this book. I am grateful to Terence Yorks for technical assistance with the pedigree and timeline figures and for his thorough preparation of the index.

DARWIN AND HIS CHILDREN

INTRODUCTION

Darwinophiles, such as me, cannot get enough books about the life and times of their hero. Our appetite for all things Darwinian is voracious. The celebration of the bicentennial year of Darwin's birth in 2009 yielded a plethora of new biographies and works on such topics as the voyages of his friends and correspondents Joseph Dalton Hooker, Thomas Henry Huxley, and Alfred Russel Wallace (McCalman 2009); Darwin's influence on photography (Prodger 2009); a biography of his wife, Emma (Loy and Loy 2010); an analysis of the illustrations in Darwin's books (Voss 2010); and even a book on Darwin's 15 or so dogs (Townshend 2009), to name just a few. In the biographies, his children, with the exception of Annie, are treated more or less as footnotes to his life, and after Charles's funeral, the accounts of his children's lives cease. The more I read about Darwin, the more I wanted to know about his family. Who were these people and what did they do with their lives? This book is an attempt to answer those questions succinctly. These sketches are based on secondary sources and are not complete biographies. That remains for others, more patient than I, to do. Nonetheless, this book assembles a considerable quantity of interesting information and illustrations in one place. It is what results when an ichthyologist gets curious about things beyond his pay grade.

The first of the 10 Darwin children, William, became a banker and managed the business affairs of the extended Darwin-Wedgwood family quite successfully. Anne died young, and Charles never quite got over it. Mary only lived a few weeks. Henrietta overcame many illnesses, and considerable hypochondria, to become her father's editor and her mother's biographer. George was a mathematician and developed into a world authority on tides. Elizabeth never married and lived with her mother and father until their deaths. She is the least known of the surviving siblings. Francis (with his father) practically invented the field of plant hormones and was a very influential plant physiologist in his own right. He made his father's autobiography and letters available to the world. Leonard had three careers: a military man, an economist/politician, and a eugenics advocate. He was a member of Parliament. Horace founded the Cambridge Scientific Instrument Company and served as Mayor of Cambridge. The last, Charles Waring, was a "special needs" child and only lived for 18 months. George, Francis, and Horace were knighted and elected Fellows of the Royal Society. The lives and accomplishments of Francis and Horace have been chronicled in detail by Ayres (2008) and Cattermole and Wolfe (1987), respectively. George and Leonard also deserve full-blown biographies of their own, and I hope this book can help stimulate that urge in a biographer looking for a subject.

What follows is a brief sketch of Darwin's life and his significance to Western culture, a chapter on his marriage to and later life with Emma, and then one chapter for each child. The biographical sketches begin with the birth of the child and end with his or her death. The time frame then starts anew with the next child. Some events in the narrative, such as Darwin's death, occur in several children's chapters. The chapters can be perused independently and in any order. My wish is that the reader comes away from this book with admiration for the accomplishments of the children of this extraordinary man, appreciation for the mutual love and devotion shown by the two parents and their children, and an understanding of the humanity of Charles Darwin.

DARWIN'S PARADIGM SHIFT

CHARLES DARWIN (1809–1882) WAS AN extraordinary man by any standard. The theory of evolution by natural selection, as elaborated in his book *On the Origin of Species* (1859) (Figure 1.1), is considered by historians and philosophers of science to be one of the most important and far-reaching ideas ever had by the human mind (Dennett 1995). Before exploring this grandiose statement, a brief review of Darwin's life and scientific accomplishments is in order. Then I will address the implications of his very useful insights that extend beyond science and profoundly impact the progress of humanity.

AN OUTLINE OF DARWIN'S LIFE

Charles Darwin was born into a wealthy English family on 12 February 1809. His father, Robert Waring Darwin (1766–1848), was a prominent physician, as was his grandfather Erasmus Darwin (1731–1802). His mother was Susannah Wedgwood (1765–1817), the daughter of Josiah Wedgwood (1730–1795), the pottery manufacturer and entrepreneur. Josiah was also a close friend of Erasmus Darwin.

Darwin's father sent Charles to medical school at Edinburgh University in 1825 and removed him in 1827, when it became obvious that Charles was not interested in a medical career. Robert Darwin then decided that Charles should study to be a clergyman in the Church of England, sending him to

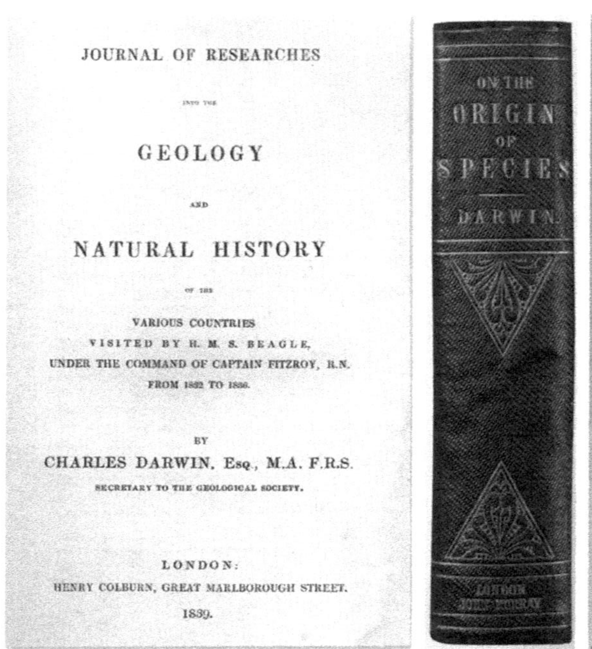

FIGURE 1.1 *Left*: The title page of Darwin's *Journal of Researches*, published in its own right as a stand-alone book. It was originally published as volume 3 of Captain FitzRoy's narrative. *The Voyage of the* Beagle was first used as the title in 1905. *Center*: The spine of the first edition of *On the Origin of Species*, by Charles Darwin, published by John Murray of London on 24 November 1859. All 1,500 copies of the first printing were ordered before the official date of publication. *Right*: The title page of the first edition of *The Variation of Animals and Plants*. Herbert Spencer's phrase "survival of the fittest" is used by Darwin for the first time here. The material included in this book extends the first chapter of *The Origin*.

Cambridge University in 1828. In 1831 Charles graduated tenth in his class among those who did not take an honors degree. He then received an invitation orchestrated by his professor, John Stevens Henslow (1796–1861), to be an unpaid naturalist-companion to Captain Robert FitzRoy (1805–1865) on a surveying voyage around the world on the H.M.S. *Beagle* (1831–1836). Darwin later described this opportunity as "the first real training or education of my mind."

On his return from the nearly five-year *Beagle* voyage, Darwin found that he was accepted as a serious scientist, and he had no desire to become a clergyman. He began working on the specimens collected during the voyage. He married his first cousin, Emma Wedgwood (1808–1896), and they eventually moved from London to Down House in Kent. They had 10 children, seven of whom survived to adulthood. In the years after the voyage, Charles was often ill, but nevertheless highly productive. He entered his ideas about how species form in a series of notebooks. This included a branching, treelike diagram that reflected the common origin and relatedness of organisms. This early evolutionary tree showed that classification should be genealogical (Pietsch 2012; Mindell 2013).

This tree's implications, however, extended far beyond taxonomy. Darwin kept his revolutionary ideas private for 20 years, except to broach them to his closest scientific colleagues: geologist Charles Lyell (1797–1875), botanist Joseph Dalton Hooker (1817–1911), and his American botanist correspondent at Harvard University, Asa Gray (1810–1888) (Porter 1993). In 1858 Darwin received a letter from naturalist Alfred Russel Wallace (1823–1913), who, like Darwin, was inspired by the writings of Thomas Malthus (1766–1834). Wallace outlined ideas on natural selection and speciation that were nearly identical to Darwin's. This letter, combined with urging from Lyell and Hooker, prompted him to complete and publish *On the Origin of Species* in 1859. Darwin continued to perform experiments and publish on a variety of topics right up to the time of his death, of heart disease, on 19 April 1882. He was laid to rest with pomp and ceremony in Westminster Abbey, a few feet from Isaac Newton. Further details can be pursued in two of the most comprehensive biographies of Darwin (Desmond and Moore 1991; Browne 1995, 2002), a concise biography (Berra 2009), and, of course, Darwin's autobiography (Barlow 1958).

SYNOPSIS OF DARWIN'S SCIENTIFIC ACHIEVEMENTS

Educated citizens are generally aware of *On the Origin of Species*, as well as Darwin's account of his voyage around the world in the H.M.S. *Beagle* through his *Journal of Researches* (1839) (Figure 1.1), a book now universally known as *The Voyage of the Beagle*. *The Voyage* and *The Origin* have never been out of print. Almost all of Darwin's books have been translated into multiple foreign languages, numbering 33 by last count (Freeman 1977). *The Origin* itself has been published in at least 29 languages, 11 in Darwin's lifetime. Only his work on barnacles appears solely in English.

Most people are surprised to learn that Darwin also made many other major contributions to geology, zoology, and botany through his observations, experiments, and writings. His books have been chronicled (Berra 2009), so I will just briefly outline the breadth of his influence. Darwin explained how coral reefs form (1842) and contributed to geological observations on movements within the earth (1844) and the deformation theory of metamorphic rock (1846). In a pioneering four-volume work that took eight years to complete, he described all known fossil and living barnacle species (1851–1854). Darwin explained how orchids are pollinated by insects (1862) and how plants climb (1865), and he catalogued the bewildering amount of variation in domestic plants and animals (1868) (Figure 1.1). He delineated human origins and sexual selection (a special form of natural selection) in multiple species in ways never before articulated (1870–1871), and discussed human and animal emotions in the same terms (1872). The latter work was one of the first books to use photographs to illustrate a point (Prodger 2009).

Darwin showed how insectivorous plants on poor-quality soils utilize nitrogen-rich insects to provide that essential nutrient (1875), and demonstrated that the offspring of cross-pollinated plants were more numerous and vigorous than self-pollinated ones (1876, 1877). His observations of growth within plants laid the foundation for the field of plant growth hormones (1880). His work on earthworms (1881) is a classic study in ecology. Any one of these achievements, by itself, could constitute a life's work for most scientists.

DARWIN'S LEGACY

Darwin was born and educated at a time when "special creation" was the prevailing scientific view. That is, God created the universe and all species a few thousand years ago, and they were unchangeable. "Revelation," not research, provided this view. Darwin began the H.M.S. *Beagle* voyage with this belief. During his lifetime the age of the earth was increasingly recognized as much more ancient, a concept suggested by James Hutton (1769–1797), Georges Cuvier (1769–1832), and Charles Lyell (1797–1875) (Bowler 1984; Larson 2004). Observations made during the voyage led Darwin to question the Genesis creation myth and the immutability of species. He found marine fossils thousands of feet above sea level and reasoned that the land had been elevated by movements within the earth, not inundated in a great biblical flood. The fossil mammals he uncovered in South America resembled living mammals from the same area. He wondered why this should be if each species was specially created. Extinction was barely recognized in those days. If each species was created in place, why did the animals on islands off continental areas resemble those on the nearest landmass? Why were there so many species in an island group that looked very similar but had slight differences from island to island? In *The Voyage of the Beagle*, Darwin concluded that it was as if "one species had been taken and modified for different ends." None of these things made sense from a creationist perspective. As he wrote to Hooker in 1844, "I am almost convinced (quite contrary to the opinion I started with) that species are not (it is like confessing a murder) immutable" (Burkhardt et al. 1985–, 3: 2).

The elegant simplicity of Darwin's reasoning can be distilled as follows. There is variation in nature, and many more offspring are generated than can survive; therefore there is a struggle for life in which favorable variations are preserved and unfavorable variations are removed. This leads to evolution, which he defined as "descent with modification," and to the formation of new species. Nature is doing the selecting for the forms best adapted to a particular environment, so Darwin called the process natural selection—as opposed to the artificial selection that breeders impose. We now know that mutation, chromosomal rearrangements, the indiscriminateness of sexual reproduction, and the like are the sources of genetic variation, but Darwin had no knowledge of such topics. Today we can speak of the descent with modification of organisms as a change in gene frequency within populations; natural selection is simply the differential reproduction of heritable traits, that is, one genetic variant leaving more offspring than another

(Berra 1990). Darwin borrowed the expression "survival of the fittest" from economist/philosopher/sociologist Herbert Spencer (1820–1903)—who published it in 1864—as a substitute for natural selection. Evolutionary fitness means reproductive fitness. In modern terms, the fittest is the one most likely to pass on the most genes to the next generation, not necessarily the biggest or the strongest individual.

By the time of Darwin's death in 1882, most scientists throughout the world had accepted the concept of common descent, but some were still skeptical of natural selection as a creative mechanism (Bowler 1984). The public was less accepting.

When the first printing of *On the Origin of Species* appeared on 24 November 1859, it precipitated one of those rare events in the history of science: a paradigm shift. Philosopher Thomas Kuhn (1962) used this term to refer to the replacement of one world view by another. Examples of a paradigm shift in science include the replacement of the earth-centered Ptolemaic system by the sun-centered Copernican system, and of Newtonian physics by relativity and quantum physics.

Darwin's work neatly dovetailed into the wider pattern of scientific advances that were occurring during his lifetime. Lyell and others had provided the necessary geological time frame for evolution to operate. The writings of Georges Cuvier, Thomas Malthus, Robert Chambers (1802–1871), Herbert Spencer, Alfred Russel Wallace, and many others helped set the evolutionary stage. By 1859 evolution by natural selection was an idea ready to burst forth. Darwin and the publication of *The Origin* made it happen. Darwin, through *The Origin* and his books that followed, changed the way humans view their place in nature. He showed that humans were not above nature, but a part of it. He supplied the explanation for the great diversity of life and showed that all life—including human—is related by descent from a common ancestor. His explanation of evolution via natural selection is the basis for all of biology and its applied subdisciplines of medicine, agriculture, and biotechnology. No other biologist in the history of our species has had an impact of this magnitude. In the words of the eminent geneticist Theodosius Dobzhansky (1973), "Nothing in biology makes sense except in the light of evolution."

The paradigm shift from creation to evolution has moved intellectual endeavors from untestable beliefs to rational understandings that flow from the scientific method. This, in turn, has allowed a vast array of advances in knowledge.

DARWINIAN IMPLICATIONS

One of the attributes of a powerful scientific theory is that it enables future research and understanding. Darwinian (or evolutionary) medicine, as formulated by Nesse and Williams (1996) and expanded by Stearns and Koella (2008) and Gluckman et al. (2009), explains how some disease symptoms, such as fever, may be a response favored by natural selection as a defense against pathogens. Some conditions generally considered to be genetic diseases, such as sickle cell anemia, may allow differential survival of its victims in malarial zones, a phenomenon called balanced polymorphism (Berra 1990). Evolutionary thinking explains the arms race waged by pathogens and hosts that prevents either from being completely eliminated. The development of resistant bacteria through the flagrant overuse of antibiotics is easily explained by Darwinian reasoning. A drug will kill the susceptible bacteria, but bacteria with a preexisting resistant mutation are unaffected and can build up the next generation. Then, when that antibiotic is later needed for a bacterial infection, the drug is ineffective. This is evolution, pure and simple.

A similar process occurs in agriculture with the overapplication of pesticides and the formation of pesticide-resistant pathogens, insects, and noxious plants. Australians are very familiar with the warfare between myxomatosis and rabbits: the virus initially killed 99 percent of the rabbits (an invasive, or nonnative, species in Australia), but, given enough time, the surviving rabbits returned in force, since the virus evolved in the direction of less virulence and the process of natural selection among the rabbits resulted in more resistance to the virus (Berra 1998).

Evolutionary psychology and evolutionary ethics, as explored by Barkow et al. (1992) and popularized by Wright (1994), help explain the origin of morality. Peacemaking among nonhuman primates, through the calming effect of mutual grooming to diffuse aggression, may be seen as the precursor of what became morality in humans (de Waal 1989). Modern religions are recent human inventions—a mere few thousand years old. The antecedents of morality, on the other hand, clearly evolved before humanity, as reflected in the empathy exhibited by bonobos (*Pan paniscus*) and the reciprocity of chimpanzees (*P. troglodytes*) (de Waal 2005). The awareness and sensitivity demonstrated by humans' closest relatives may be the underlying driver of prosocial behavior (de Waal 2012, 2013). Kin selection, where an individual voluntarily sacrifices for a close genetic relative, makes sense in an evolutionary context, because some of the same genes of the individual making the sacrifice will be passed on by the kin who survives. Hamilton (1972) refers to this as inclusive fitness. A realization that humans share kinship with all animal life has helped to raise consciousness about how we treat other animals (Singer 1977).

The ancestry of the AIDS virus, HIV-1 (human immunodeficiency virus-1), has been traced to SIVcpz (simian immunodeficiency virus) carried by our closest living relative, the chimpanzees, *P. troglodytes* (Bailes et al. 2003). This is not surprising from an evolutionary perspective. Somewhere in high school today there is a student whose future research may contribute to better control of the AIDS epidemic. What chance of that would there be if evolution weren't taught properly in high school?

Even religion is now being explained as having an evolutionary origin: a natural phenomenon that arose once the brain evolved a critical mass and complexity (Dennett 2006). Bloch (2008) suggested that the evolution of imagination was a requisite for the emergence of religion, which he considered a logical extension of human sociality. Previously, this emergence of modern human behavior was thought to have occurred about 35,000–40,000 years ago, the time of the Upper Paleolithic "revolution," as manifested by an explosion of image making and cultural transformations (White 2003). However, recent discoveries in South Africa of engraved ochres (Henshilwood et al. 2002) and of small bladelets made from heat-treated stone (K. Brown et al. 2012) demonstrate that humans had already evolved the capacity for complex thought at least 70,000 years ago. Acceptance of authority (necessary for group cohesion and survival), enforced by tool use and language, and combined with a confusion between coincidence and cause and effect, can result in the establishment of a religious belief that becomes dominant in a culture (Wolpert 2007). Religion encourages beliefs and rituals—which may appear absurd to outsiders—that unite in-group cohesiveness but also promote conflict with out-groups (Atran and Ginges 2012).

Those whose religion requires a literal interpretation of the Bible fear that a paradigm shift from supernaturalism to methodological naturalism (a naturalistic causation for nature's phenomena) threatens their beliefs. The 1925 Scopes trial—nicknamed the "monkey trial" and the "trial of the century"—in Dayton, Tennessee, has come to symbolize the struggle of religion against science in popular culture; the trial later inspired the play and film *Inherit the Wind* (Larson 1997). Such creationists are particularly vocal in America, which has a longstanding tradition of anti-intellectualism (Pigliucci 2002; Numbers 2006). This has resulted in a series of creationist legal challenges to evolution that have been decided in favor of evolution (Berra 1990). The most important legal cases are against creationists or government entities that have adopted creationist policies, thus violating part of the US Constitution. These legal decisions include *Epperson v. Arkansas*, *McLean v. Arkansas*, *Edwards v. Aguillard*, and *Kitzmiller v. Dover*. In the latter case, intelligent design (ID) creationists influenced the Dover, Pennsylvania, School Board to adopt their ideas, an action that was challenged in the courts by a group of parents. The ID creationist philosophy, which posits that life is too complex to have arisen by natural means and therefore had a supernatural origin, has been critiqued in Pennock (2001) and exposed as a threat to science education by Forrest and Gross (2004). In the concluding portion of his decision, Judge John E. Jones III (2005) determined that the school board's policy of teaching intelligent design violated the Establishment Clause (the separation of church and state) of the First Amendment to the US Constitution. He wrote: "In making this determination, we have addressed the seminal question whether ID is science. We have concluded that it is not, and moreover that ID cannot uncouple itself from its creationist, and thus religious, antecedents.... The breathtaking inanity of the board's decision is evident when considered against the factual backdrop which has now been fully revealed through this trial." For those who want to dive deeper into the miracle-strewn world of the anti-science crowd and explore this interesting case further, Padian (2007) reviewed three books based on the Dover trial.

MODERN EVOLUTIONARY SYNTHESIS

Darwin, of course, had no knowledge of genes, chromosomes, or how inheritance worked. This required additional input, arising from an understanding of Gregor Mendel's (1822–1884) genetic work. Biotechnology, whether in the form of genetically modified crops, designer drugs, gene therapy, or the human genome project, derives from Darwin's and Mendel's profound insights into how nature operates.

The modern evolutionary synthesis grew from Darwin's explanation of natural selection and Mendel's demonstration that inheritance was particulate—that is, that it can be passed from generation to generation by "particles," now known as genes (Dobzhansky 1937)—augmented by the research of mathematically oriented population geneticists such as J. B. S. Haldane, Ronald A. Fisher, Sewall Wright, Thomas Hunt Morgan, Theodosius Dobzhansky; paleontologist George Gaylord Simpson; botanist G. Ledyard Stebbins Jr.; biologist Julian Huxley (Thomas Henry Huxley's grandson); and the most important evolutionary biologist since Darwin, Ernst Mayr. This fusion of knowledge moved evolutionary science forward to the middle of the twentieth century (Larson 2004). James D. Watson and Francis Crick's 1953 demonstration that the molecular structure of DNA (deoxyribonucleic acid) allowed for genetic coding was a huge breakthrough, one that ultimately made it possible to sequence the three billion chemical base pairs that compose the human genome and identify the approximately 20,000–25,000 genes in human DNA (Lander et al. 2001; Venter et al. 2001).

TV viewers are familiar with DNA analysis, popularized on various CSI (crime scene investigation) programs. DNA-sequencing techniques—where the arrangements of the A-T-C-G nucleotides are compared—can convict or exonerate people accused of crimes. Similar techniques can confirm or deny paternity in disputed cases, or can ensure that the expensive grouper fillets you purchase are not flesh from lesser species. Such evolutionary tests are accepted by the judicial system because they pass the Daubert standard for scientific evidence: the techniques were subject to empirical testing, published in peer-reviewed journals, and accepted by the scientific community (Mindell 2009).

Recent discoveries in evolutionary developmental biology, known as evo-devo, have shown that very similar genes are present in very dissimilar animals. These body-shaping genes are controlled by DNA switches (called enhancers) that turn them on or off at various times during development. Such enhancers are a major factor in the development of morphology, the branch of biology that deals with the form of living organisms and with relationships between their structure (Carroll 2005). The above examples are just a smattering of the benefits to society that flow directly from the creative power of Charles Darwin's theory of evolution by means of natural selection.

The Human Genome Project spawned ENCODE (Encyclopedia of DNA Elements), whose mission was to describe all of the functional elements encoded in our genome. The 6 September 2012 issue of *Nature* published six coordinated ENCODE papers, while 24 related papers were published elsewhere that same week. (Further exploration of this complicated, state-of-the-art topic is facilitated at www.nature.com/encode/.) In addition to well-known coding elements, ENCODE explained the hidden genetic switches that regulate development and turn other genes on and off. This new knowledge showed that the term "junk DNA" was just a manifestation of our incomplete understanding. Today's biologists are fortunate to have the very broad shoulders of Charles Darwin to support and make possible their elaboration of how biology works.

The paradigm shift (Berra 2008) instigated by Darwin has made more obvious the superiority of the scientific method as a means of understanding the world around us. It is ironic that the legacy of a man once destined for the church has been to replace supernaturalism with methodological naturalism.

2

CHARLES AND EMMA

After four years, nine months, and five days at sea, the H.M.S. *Beagle* returned to England on 2 October 1836. Charles Darwin was 27 years old. He immediately visited with his father, Dr. Robert Waring Darwin, and his sisters at the family home in Shrewsbury on 4 October. By 15 October he went to Cambridge University to see his mentor, Professor John Stevens Henslow, where he set about analyzing the specimens he collected on the voyage around the world and began writing up his notes for publication. Thanks to Henslow, by the time Darwin returned from his voyage, he was already a well-known scientist, much to his surprise. Henslow had shown Darwin's letters and geological descriptions to fellow scientists and had displayed some of the specimens Darwin had shipped to him from South America. These wonders had caused quite a stir in scientific circles (Desmond and Moore 1991).

LIFE AFTER THE H.M.S. *BEAGLE*

Darwin traveled to London to see his brother, Erasmus Alvey Darwin, on 20 October. He also visited Maer Hall, located near Staffordshire, seven miles from Stoke-on-Trent, on 12 November. Maer Hall was the estate of his uncle, Josiah Wedgwood II, and the home of a multitude of his cousins. Charles was back in London by 2 December, where he met Charles Lyell, the president of the Geological Society, and consulted with Richard Owen about the fossils he (Darwin) had collected. He attended meetings of the Geological Society and the Cambridge Philosophical Society, and he generally entered into the society scene, including the Athenaeum Club. Darwin presented a paper on the geological uplift of the coast of Chile at the 4 January 1837 meeting of the Geological Society, with Lyell in attendance (Browne 1995).

Charles took up residence in London at 36 Great Marlborough Street on 13 March 1837, in order to be near his brother. Erasmus had earned a medical degree at Edinburgh University, but he never practiced medicine and lived as a socialite on his allowance from their father. Charles continued to work on his *Journal of Researches*, based on his H.M.S. *Beagle* diary, and attended scientific meetings, where he presented papers on extinct mammals, Galápagos finches, and coral reefs.

Charles had been trying to make sense of his observations during the H.M.S. *Beagle* voyage about how species vary from place to place—from mainland to island. In July 1837 he opened his first notebook on the transmutation of species, and in notebook B he drew his first tree of life, showing the common ancestry of all living things. He would eventually fill 15 notebooks with his thoughts on geology, the transmutation of species, and metaphysical ideas (Barrett et al. 1987). In September he experienced heart palpitations, which were to dog him throughout his life, especially during times of stress and overwork. Charles was elected a secretary of the Geological Society on 16 February 1838 and served in that capacity until 1841.

MARRIAGE TO EMMA WEDGWOOD

Charles's older sister, Caroline Sarah Darwin, married her first cousin Josiah Wedgwood III in 1837. Charles, too, was beginning to entertain thoughts of settling down. On the back of a note from geologist and linen-draper Leonard Horner, written in April 1838 (Browne 1995), Charles listed the pros and cons of marriage. The balance came down in favor of marriage.

At the age of 29 Charles decided it was time to take the plunge. He consulted his father, who advised Charles not to discuss the Darwin family's lack of religious belief with a prospective wife, because "some women suffered miserably by doubting about the salvation of their husbands, thus making them likewise to suffer" (R. Keynes 2001). His thoughts turned to his first cousin, Emma Wedgwood, whom he had known since childhood. She had all the qualities he desired in a wife. Charles's mother was a Wedgwood, as were his maternal grandmother and grandfather (Berra et al. 2010a, 2010b). He knew what he was getting into. Emma's father was his beloved uncle, Josiah Wedgwood II, the man who interceded with his father to allow Charles to go on the voyage of the H.M.S. *Beagle.*

Meanwhile, in October 1838, Charles read Thomas Malthus's book, *An Essay on the Principle of Population.* Inspired by Malthus's thinking, in his autobiography Charles wrote: "It at once struck me that under these circumstances favorable variations would tend to be preserved, and unfavorable ones destroyed. The result of this would be the formation of new species. Here, then, I had at last got a theory by which to work."

Refocusing on his personal life, Charles proposed marriage to Emma at Maer Hall on Sunday, 11 November 1838, and she accepted immediately. Charles did not heed his father's wise council and candidly discussed with Emma his lack of belief in things that could not be proven scientifically. Before their marriage, Emma, a devout Unitarian, wrote to Charles, "My reason tells me that honest and conscientious doubts cannot be a sin, but I feel it would be a painful void between us" (Burkhardt et al. 1985–, 2: 122–123; Healey 2001).

Emma was undeterred, however, and accepted Charles as he was: a sincere, honest, and kind man. In early December they went house hunting in London. On 31 December, Charles, with the help of his servant from the H.M.S. *Beagle*, Syms Covington, took possession of the house at 12 Upper Gower Street. It was located in a well-to-do professional neighborhood, near University College and its teaching hospital. Emma's brother, Hensleigh Wedgwood, and his wife Fanny lived a couple of doors away (Browne 1995). Surgeons, artists, and lawyers populated the area. Charles dubbed this house Macaw Cottage because of its gaudy yellow curtains, blue walls, and multicolored furniture.

On 24 January 1839 Charles was elected a Fellow of the Royal Society, the highest honor in British science. Charles and Emma were married on 29 January 1839 in the little church at Maer Hall (Figure 2.1). Charles was two weeks shy of turning 30 years old, and Emma was 30 years and 8 months old. They began their married life with a substantial financial cushion from their wealthy families. Dr. Robert Darwin provided investments of £10,000 that would yield about £600 in annual dividends. In addition, Josiah Wedgwood II gave them £5,000 and a £400 annual allowance. In modern terms, their wealth was the equivalent of a $1,250,000 portfolio that could be expected to yield approximately $83,000 annually (Loy and Loy 2010). One of Emma's first purchases for their new house was a pianoforte, which she would lovingly play for Charles each evening. Visitors to Macaw Cottage included Dr. Henry Holland and his wife, Charles Lyell and his wife, Thomas and Jane Carlyle, anatomist Richard Owen, and writer Harriet Martineau, a friend of Charles's brother Erasmus. The Darwins attended the local Unitarian chapel with Hensleigh and Fanny Wedgwood.

FIGURE 2.1 Emma Wedgwood, in an 1840 colored chalk painting by George Richmond, commissioned by her father, Josiah Wedgwood II. Charles Darwin in a watercolor painting, also by George Richmond, in 1840. Darwin Heirlooms Trust.

EMMA'S FEELINGS

In April 1839 Emma realized that she was pregnant. Perhaps while pondering the very real possibility of death during childbirth, Emma again wrote to Charles about the thoughts she could not express to his face: "Everything that concerns you concerns me and I should be most unhappy if I thought we did not belong to each other for ever." She declared her happiness and love and acknowledged his affection, "which makes the happiness of my life more and more every day." Charles sentimentally wrote on the outside of this note, probably many years later, "When I am dead, know that many times I have kissed and cried over this.

C. D." (Burkhardt et al. 1985–, 2: 172). Years later, Charles wrote the following for his children in his *Autobiography*:

> You all know well your Mother, and what a good Mother she has ever been to all of you. She has been my greatest blessing, and I can declare that in my whole life I have never heard her utter one word which I had rather have been unsaid. She has never failed in the kindest sympathy towards me, and has borne with the utmost patience my frequent complaints from ill-health and discomfort. I do not believe she has ever missed an opportunity of doing a kind action to anyone near her. I marvel at my good fortune that she, so infinitely my superior in every single moral quality, consented to be my wife. She has been my wise advisor and cheerful comforter throughout life, which without her would have been during a very long period a miserable one from ill-health. She has earned the love and admiration of every soul near her.

WILLIAM ERASMUS DARWIN

(1839–1914)

On 15 August 1839, Charles's first book, *Journal of Researches into the Geology and Natural History of Various Countries Visited by H.M.S.* Beagle *Round the World under the Command of Capt. FitzRoy, R.N.* was published in its own right. Previously it was published as volume 3 of Captain FitzRoy's *Narrative*. The *Journal of Researches* is now known as *The Voyage of the Beagle.*

FIRST CHILD

Charles and Emma's first child, William Erasmus Darwin, was born on 27 December 1839. Emma's unmarried and oldest sister, Sarah Elizabeth ("Bessy") Wedgwood, arrived to help with the delivery. These were the days before chloroform became available as an anesthetic, and Charles reported that the event "knocked me up, almost as much as it did Emma herself" (Burkhardt et al. 1985–, 2: 270). Emma's brother, Hensleigh Wedgwood, and his wife Fanny lived only four doors away and were also popping in and out at all hours to help Emma. Emma regained her strength by February 1840 (Healey 2001). In the next 17 years, Emma was to give birth nine more times!

THE NATURAL HISTORY OF BABIES

Charles referred to his son as "a little prince" (Burkhardt et al. 1985–, 2: 250), and he wrote to Captain FitzRoy about "my little animalcule of a son." As an infant, William was nicknamed "Hoddy Doddy" and, later, "Willy" (Freeman 1978). Ever the observant scientist, Charles immediately began making notes on William's expressions, reflexes, and general behavior. During the first seven days after birth, he recorded William's sneezing, hiccupping, yawning, stretching, sucking, and screaming, as well as his reaction to tickling. These observations continued into 1841 and were the subject of a scientific paper in the quarterly journal *Mind* (C. Darwin 1877b). Basically, Charles was gathering data on the natural history of babies. In Darwin's groundbreaking book, *The Expression of Emotions in Man and Animals*, he discussed tear production in infants, based on his observations of William (C. Darwin 1872). This tome was one of the first books to use photography to illustrate the text (Prodger 2009).

A few passages from his paper in *Mind* give one a feel for Darwin's observations of William's behavior. "At the age of 32 days he perceived his mother's bosom when three or four inches from it, as was shown by the protrusion of his lips and eyes becoming fixed." "When 77 days old, he took the sucking bottle... with his right hand, whether he was held on the left or right arm of his nurse, and he would not take it in his left hand until a week later although I tried to make him do so; so that the right hand was a week in advance of the left. Yet this infant afterwards proved to be left-handed." "When he was 2 years and 4 months old, he held pencils, pens, and other objects far less neatly and efficiently than did his sister [Anne Elizabeth] who was then only 14 months old, and who showed great inherent aptitude in handling anything."

Darwin, the loving father, is reflected in this passage: "The first sign of moral sense was noticed at the age of nearly

13 months. I said 'Doddy won't give poor papa a kiss,—naughty Doddy.' These words, without doub't, made him feel slightly uncomfortable; and at last when I had returned to my chair, he protruded his lips as a sign that he was ready to kiss me; and he then shook his hand in an angry manner until I came and received the kiss." When William was two years and eight months old, Darwin described his devious behavior. Apparently young William had secreted a pickle in his pinafore and was acting suspiciously. William "repeatedly commanded me to 'go away', and I found it [the pinafore] stained with pickle juice; so that here was carefully planned deceit. As the child was educated solely by working on his good feelings, he soon became as truthful, open, and tender, as anyone could desire."

In March 1838 Darwin made comparative observations on a young orangutan at the London Zoo. Three-year-old Jenny was the first orangutan to be displayed at the zoo and caused a sensation with the public at large (Desmond and Moore 1991). Darwin was struck by her human, naughty emotions, much like those of a child. He compared Jenny's reaction to a mirror with William's more complicated response.

William suffered the indignities of both baptism and vaccination against smallpox. Charles obsessed over his own poor health and that of his children (Berra et al. 2010a, 2010b), but William was the one child that appeared robust and devoid of the hypochondria so prevalent in the rest of the Darwin household.

SCHOOL DAYS

An 1842 daguerreotype of 33-year-old Charles with approximately three-year-old William seated on his lap is the only known photograph of Darwin with a family member (Desmond and Moore 1991) (Figure 3.1). As a 10-year-old boy, William had an interest in collecting butterflies (Figure 3.2). At about that age he

FIGURE 3.1 Charles Darwin and his first child, William, in an 1842 daguerreotype, the only known photograph of Charles with a family member. Library, University College, London.

was also given a pony, which Charles taught him to ride. William was initially educated at the Reverend Henry Wharton's preparatory school at Mitcham and then, in February 1852, entered Rugby School in Warwickshire, the genteel, Anglican school of his Wedgwood uncles. Charles felt that the Rugby School experience closed rather than opened Willy's mind (Healey 2001). William injured his ankle playing football at that school and wore an iron support for many years (F. Darwin 1914). Charles missed his son when the boy was away at school and ensured that Willy would spend summers at Down House with the family, including Uncle Ras (Charles's brother, Erasmus), who was a frequent guest.

William developed an interest in photography and received a yearly allowance of £40 to pay for supplies. This allowance

FIGURE 3.2 Daguerreotype of nine-year-old William Erasmus Darwin, from January 1849, by Claudet of London. Darwin Museum, Down House.

was almost as much as Joseph Parslow received annually as the family's beloved longtime butler (Freeman 1978). Darwin encouraged William to take photographs around the house and garden (Figure 3.3). An upstairs room was converted into a darkroom, where William most likely used a wet-plate collodion process to make negatives by hand (Prodger 2009). This was a very demanding technique, even for professional photographers.

CAMBRIDGE UNIVERSITY

In 1858 William entered Christ's College at Cambridge University and eventually resided in the same room his father

FIGURE 3.3 *Left*: Photograph of Charles Darwin, taken by William Erasmus Darwin, at about the time of the publication of *The Origin*. Charles did not grow a beard until 1862, when shaving became too irritating for his eczema. Emma was the one who proposed this solution (Burkhardt et al. 1985–, 10: 627; Colp 2008). Gray Herbarium Archives, Harvard University. *Right*: William Erasmus Darwin, photographed by S. J. Wiseman, Southampton. Note William's resemblance to his father. English Heritage Photo Laboratory.

once occupied (Figure 3.4). This room is now distinguished by a commemorative Wedgwood plaque and inscription (Freeman 1978). Moreover, Charles Darwin's Cambridge lodging is alleged to be the same room previously occupied by William Paley, the theologian and author of *Natural Theology*, published in 1802 (Browne 1995). Paley reasoned that if you find a watch, there must be a watchmaker. This is the basis for the argument made by intelligent design creationists today. As a theology student at Cambridge, Charles Darwin read Paley's work and wrote in his *Autobiography* that "*Natural Theology* gave me as much delight as did Euclid." Once he discovered natural selection, however, Darwin realized that the apparent design in nature is really the undirected operation of natural forces that give the illusion of

FIGURE 3.4 One of Charles Darwin's rooms at Christ's College, Cambridge, as it looked in 1909. This photograph by J. Palmer Clark, to mark the 100th anniversary of Darwin's birth, was published in the *College Magazine* for 1909. Charles's quarters were restored to the appearance they had during Darwin's tenure for the 2009 bicentenary of his birth. *Inset*: William Erasmus Darwin at age 22, who also occupied the same Cambridge room, photographed by William Farren, about 1861. Darwin Museum, Down House.

design. This was forcefully explained by Dawkins (1986) in the aptly titled *The Blind Watchmaker*.

According to Van Wyhe (2009), no records exist to authenticate that Paley and Darwin were assigned the same room. He described the story as a "tradition" at Christ's College, but added that "many such college traditions are surprisingly accurate." One can only hope that such a symmetrical tale is true.

William was on the rowing team at Cambridge from 1859 to 1861. His crewmates regarded him with affection, and they remarked on the enthusiasm and effectiveness of his rowing (F. Darwin 1914). As he grew bald later in life, William used his long-treasured blue silk nightcap (the colors that distinguished his Christ's College boat from others), to cover his bald spot (F. Darwin 1914).

WILLIAM THE BANKER

Charles envisioned his son as a barrister and eventually as "Lord Chancellor of all England" (Desmond and Moore 1991). William studied mathematics and gave up the idea of becoming a barrister when offered a position as a partner in Grant and Maddison's Bank in Southampton. Sidestepping the old-boy banking network was facilitated by a Darwin neighbor, John Lubbock, but Charles had to negotiate a £5,000 partners' guarantee for a capital reserve in the event of a run on the privately owned bank. This bank eventually became Lloyds Bank.

William left Cambridge without a degree to accept the banking position, but he returned in 1862 and completed his B.A. examinations (Browne 2002). He received a B.A. degree with honors and applied for and received an M.A. in 1889 (Loy and Loy 2010). William stayed with the banking firm for 40 years, from 1862 to 1902. As a banker, he quite successfully managed the financial affairs of Charles and Emma and the extended Darwin family. In addition to the Darwin-Wedgwood wealth, Charles did very well with his books. For example, by 1861 the British and American royalties from *The Origin* totaled £1,219, the modern equivalent of about $100,000 (Loy and Loy 2010). William even stepped in to manage the accounts of Joseph Parslow, the family's retired butler, when Parslow got into financial trouble. William eventually became a trustee for the inheritance of his brothers and sisters, as well as executor for his parents, uncles, and aunts (Browne 2002). When Charles's brother Erasmus died in 1881 and his estate was settled, Charles Darwin's personal fortune exceeded

£250,000, or about $21 million in today's terms (Loy and Loy 2010). William clearly had major responsibilities entrusted to him.

HELPFUL SON

As one trained in mathematics, William reliably did the complex calculations involved in Charles's study of purple loosestrife, *Lythrum salicaria*, during the summer of 1862 (Browne 2002). Charles had been overwhelmed by the statistical possibilities of the three-way pollination patterns of this plant. William also did some drawings and dissections for his father and read proofs. In Charles's later years William acted on behalf of his father to prevent the use of Darwin's name in unwanted endorsements of hot political topics (Desmond and Moore 1991).

THE EYRE AFFAIR

Righteous indignation flared between Charles and William over the Eyre affair and the Jamaican Committee. Edward John Eyre was governor of Jamaica and brutally put down an insurrection of blacks in 1865. Over 400 local peasants were killed, 600 flogged, and 1,000 houses destroyed (Desmond and Moore 2009). This raised the ire of the antislavery Darwinians, including Charles Darwin, Alfred Russel Wallace, Charles Lyell, and Thomas Henry Huxley. Liberal groups formed a Jamaica Committee to advocate for the prosecution of Eyre, while clergymen, members of the military, peers, and even Charles's dearest friend, Joseph Dalton Hooker, rallied to Eyre's defense. When William, now 26 years old, made derogatory remarks about the Jamaica Committee at Uncle Ras's house, Charles couldn't contain his anger. He told William that if he really felt that way, William "had better go back to Southampton." Charles abhorred slavery and all forms of cruelty. Early the next morning, though, Charles sat on his son's bed and apologized for his strong outburst (Desmond and Moore 1991, 2009).

Eyre was tried three times and acquitted each time. His legal expenses were paid, and he was granted a government pension. Eyre was the first English explorer to view the huge ephemeral salt lake in South Australia in 1840. Lake Eyre was named for him in 1860.

ASA GRAY AND FRIENDS

Asa Gray (Harvard University botanist, friend, and correspondent of Darwin) and his wife visited Down House in 1868. They dined with the Darwins, the J. D. Hookers, and physicist John Tyndall. Charles had first met Gray and his wife at Kew Gardens in the spring of 1851, and he began an extensive correspondence with Gray on 25 April 1855 (Burkhardt et al. 1985–, 5: 322–323). Gray eventually became a strong supporter of Darwin's ideas on evolution in America. A mutual friend of Gray's and Darwin's, Charles Eliot Norton, was staying nearby and dropped into Down House for lunches, accompanied by his wife and her 29-year-old sister, Sara Sedgwick, who caught William's eye. This pleased Emma, who once wrote to her son advising that he choose a wife only from family friends (Browne 2002).

OULESS OIL PORTRAIT

It was William who, after several years of suggestions, finally persuaded his father to sit for an oil portrait. Artist Walter

Ouless stayed at Down House in February and finished the first oil painting of Darwin in March 1875. The Darwin family owns the original, and a copy hangs in Christ's College, Cambridge University. Francis Darwin wrote, "Mr. Ouless's portrait is, in my opinion, the finest representation of my father that has been produced." Charles said, "I look a very venerable, acute, melancholy old dog; whether I really look so I do not know" (Freeman 1978). Emma didn't like the painting, considering it rough and dismal, and she removed it from view (Browne 2002).

CHINESE GORDON

In Southampton in 1877, William became acquainted with Charles George Gordon of the Royal Engineers. Colonel, and later General, Gordon was known as "Chinese Gordon" for his service in the Orient during the Second Opium War and "Gordon of Khartoum" when he was governor general of the Sudan (Davenport-Hines 2004). He was speared and beheaded in the siege of Khartoum in 1885, and his body was never found. His exploits were immortalized in the 1966 Hollywood epic *Khartoum*, staring Charlton Heston. William, Gordon, and Gordon's sister dined together and seemed to enjoy each other's company enough to meet on other occasions (Loy and Loy 2010). Gordon regaled William with his adventures, war stories, and eccentric views on religion and free will.

MARRIED LIFE

When William and Sara Sedgwick of Cambridge, Massachusetts, the daughter of a New York lawyer and an English mother, finally got engaged, the Darwins gave them a gift of £300 (Desmond

FIGURE 3.5 William Erasmus Darwin and his sister Henrietta Darwin Litchfield. Darwin Museum, Down House.

and Moore 1991). William married Sara on 29 November 1877, nine years after they first met. Sara was 38 years old and William nearly so. William and Sara's home at Basset, Southampton, became a favorite destination for the short holidays forced by Charles's ill health (F. Darwin 1914). William's niece, Gwendolen ("Gwen") Mary née Darwin Raverat, the daughter of George Howard Darwin, described the household at Basset and, by implication, Aunt Sara, as "painfully tidy, psychologically clean, rigidly perfect in organization" (Raverat 1952). William and Sara sailed to America in September 1878 to meet Sara's family and to visit such Darwin correspondents as Asa Gray at Harvard University (Loy and Loy 2010).

William remained physically very active. As adults, William and his brother Leo bicycled the 16 miles from London to Downe (Loy and Loy 2010). A broken leg during a hunting accident, when William rashly "tried to go through a swinging gate," resulted in the amputation of his leg, but he carried on without complaint (Healey 2001). He eventually got a wooden leg and resumed an active life, although he used a cane (Figure 3.5). He read widely on art, science, history, and biography. William was an amateur geologist and botanist and made some observations on pollination that were reported in the second edition of his father's orchid book (C. Darwin, 1877, 99).

CHARLES DARWIN'S FUNERAL

When Charles died at Down House on 19 April 1882, Emma and the family, as well as the village of Downe, wanted Charles to be buried in the little church cemetery, next to his brother Erasmus. Twenty members of Parliament, led by Charles's friend and neighbor John Lubbock (who was also president of the Linnean Society) and supported by Charles's cousin Francis Galton, Thomas Henry Huxley and Joseph Dalton Hooker, as well as William Spottiswoode (president of the Royal Society), requested that Darwin be laid to rest in Westminster Abbey, with the honor befitting the national treasure he had become. William welcomed this idea, and Emma, realizing that Charles would have wanted it also, gave her permission (Browne 2002; Loy and Loy 2010). The behind-the-scenes machinations of how this was brought about were artfully described by Moore (1982). Darwin's closest scientific friends—Hooker, Huxley, Wallace, Lubbock, and Spottiswoode—as well as James Russell Lowell, American envoy extraordinary and minister plenipotentiary to the

Court of St. James (the United States did not have ambassadors until 1893), were among the 10 pallbearers.

William was now head of the family, and he led the funeral procession, with 33 Darwins and Wedgwoods following the coffin (Healey 2001). At his father's funeral in Westminster Abby on 26 April 1882, William, seated in the front row, felt a draught on his bald head. Fearing a cold, he positioned his black gloves on his head and remained that way until the end of the service (Desmond and Moore 1991). That must be what Gwen Raverat (1952) meant when she described William as the most unselfconscious of all her uncles!

The *Times* [London] (26 April 1882) described the funeral. Huxley wrote a loving obituary that appeared in *Nature* on 27 April 1882 (T. Huxley 1882). It contained the following elegant tribute: "One could not converse with Darwin without being reminded of Socrates. There was the same desire to find some one wiser than himself; the same belief in the sovereignty of reason; the same ready humour; the same sympathetic interest in all the ways of men. But instead of turning away from the problems of Nature as hopelessly insoluble, our modern philosopher devoted his whole life to attacking them in the spirit of Heraclitus and of Democritus, with results which are the substance of which their speculations were anticipatory shadows."

WIDOWER

After the death of his wife Sara in 1902, William retired from the bank and moved to London. He settled at 11 Egerton Place, next door to his brother Leonard Darwin, where he experienced a "second youth" (Raverat 1952). The Darwin family often gathered there. Gwen lived at William's house while she was an art student at the Slade School of University College, London. She married Jacques Raverat and wrote and illustrated a delightful childhood memoir of life in the Darwin household, titled *Period Piece* (Raverat 1952). This book provided many revealing anecdotes about the Darwin children (Gwen's uncles and aunts), including the charming and lovable nature of Uncle William. She wrote that "he was as nearly made of pure gold as anyone on this earth can be." When walking arm in arm with her widowed uncle, Gwen described a scene in which a pretty girl would pass them. Uncle William "would stop dead, turn around, stare, and say loudly: 'Good looking young woman that.'" She described him as having fresh pink cheeks, clear blue eyes, shaggy eyebrows, and white hair. Gwen remembered William's brother Francis (Gwen's Uncle Frank) once saying that "you could eat a mutton chop off William's face," because he was so clean and wholesome.

William was a strong supporter of university education for people of all classes, and he championed the foundation of a university college in Southampton in 1902. He owned an automobile, a White steam car, that he named Betsey. It was driven by a chauffeur, and William would take his nieces and nephews on excursions, sometimes even 30-mile trips. It frequently broke down and had to be pushed (Raverat 1952).

In a rare public appearance, William spoke lovingly of his father at a dinner given by Cambridge University for the Darwin Centennial in 1909. William died suddenly on 8 September 1914 in Sedbergh, where he was spending the summer with the family of his brother George (F. Darwin 1914). He was buried beside the grave of his wife at St. Nicholas's Church, North Stoneham, near Southampton.

ANNE ELIZABETH DARWIN

(1841–1851)

In early 1841 Darwin was still working on parts of the bird and fish sections of *The Zoology of the Voyage of the H.M.S. Beagle.* He edited and superintended this magnificent work of 632 pages and 166 plates, which was published in five parts from 1838 to 1843. Charles, Emma, and 14-month-old William were then living at 12 Upper Gower Street in London (Browne 1995). In 1842 the Darwin family moved to Down House, located in the village of Downe in Kent (see chapter 5).

ANNIE'S BIRTH

Anne arrived into the world on 2 March 1841. Once again Emma's sister, Elizabeth Wedgwood, arrived to help. Dr. Henry Holland served as obstetrician for the difficult birth. Charles, now 32 years old, immediately fell in love with his first daughter, who was to become his favorite child. He referred to her as "Kitty Kumplings" when she was an infant. Eventually everyone called her "Annie" (Freeman 1978). Charles's great-great-grandmother was Anne Waring, and his Cambridge professor, John Henslow, also had a daughter named Anne. Charles immediately began recording his developmental observations regarding Annie and comparing them with Willy's progress.

By the time she was two years old, Charles observed that she showed "no signs or skill in throwing things... as Willy did." He also added, "Nor does she so readily give slaps." Charles attributed this to the innate differences between boys and girls.

The children provided a welcome diversion from his constant work. In May 1842 Charles wrote out a brief sketch of his theory of evolution for his personal use, but he did not publish it. His book on *The Structure and Distribution of Coral Reefs*, however, was published that year (Freeman 1977).

AN IDYLLIC CHILD AND CHILDHOOD

Seven-year-old Annie curried her papa's favor by popping into his study with a pinch of snuff from a container he hid upstairs, since he was trying to avoid using it (Desmond and Moore 1991). Charles welcomed, this as he did all of Annie's displays of kindness to him. Almost everything she did endeared her to him more deeply, and he doted on her. He loved it when she combed his hair. Annie was highly empathetic, like her mother, with a sunny, cheerful temperament. She grew tall for her age, and had gray eyes and long brown hair (Desmond and Moore 1991).

Emma raised her children in the Unitarian Christian tradition. Each Sunday she took them to the local Anglican church of the Reverend John Brodie Innes, the vicar of Downe. Charles did not attend, but Annie had no idea of her father's lack of religious belief.

Annie delighted in playing with the family butler, Joseph Parslow, who had been Charles's manservant in Upper Gower Street. Parslow remained an integral part of the family and served at Down House until 1875 (Freeman 1978). He married Eliza, Emma's maid, who later opened a sewing shop in Downe. Uncle Ras (Charles's brother Erasmus) was a favorite playmate of all the Darwin children, as was Joseph Dalton Hooker, Charles's botanist friend who would stay for a week and bring work to discuss with Charles. Annie enjoyed roughhousing with her brothers and sisters in the 18 acres of lawns, gardens, meadows, and woodlands of Down House. She had her own garden plot. She also played with the gardener's children and enjoyed the various cows, pigs, horses, chickens, and wildlife on the property. Willy and Annie took strolls with their father along the Sandwalk (a walking path around the perimeter of Down House), and interacted with the neighbor children of Sir John Lubbock, a banker, scientist, and important person in the village of Downe. Charles spent a substantial amount of time tutoring Lubbock's eldest son, John, in natural history, and they became fast friends throughout Charles's life. John eventually became a member of Parliament (R. Keynes 2001).

Sir John Lubbock established a school for boys in Downe, in order to provide them with a secular education instead of religious indoctrination. Charles supported this effort by paying the tuition for some needy children. The progressive attitude of Whigs like the Lubbocks and Darwins promoted "useful knowledge" and considered religion a hindrance to such goals, because of the doctrinal arguments and sectarian feelings it generated (R. Keynes 2001).

HOME SCHOOLING

The Darwins were a family of readers, and Annie's reading list included *The Last of the Mohicans*, *Gulliver's Travels*, and *Arabian Nights* (R. Keynes 2001). The family's view of education was that the children should be encouraged to think and learn for themselves, without strict discipline and with a sense of freedom. They should learn from their experiences. Charles would not force his children to be interested in natural history, but growing up in such a household in the country, it was inevitable that they would be exposed to nature. They could choose to explore it, if they wished. In wealthy families, tradition dictated that boys were sent to boarding schools and girls were educated at home. Home schooling fell under Emma's purview, and she was rather easygoing. This reflected her progressive upbringing as a Wedgwood family member. In the summer of 1848, Emma hired 19-year-old Catherine Thorley as a governess for the children. Miss Thorley's job was to teach needlework, music, dancing, French, and etiquette. She was also engaged to turn the girls into gentlewomen. Charles paid her £50 annually. Annie and Miss Thorley spent a lot of time together and grew very close (R. Keynes 2001). The governess became an important part of the family, and she even worked with Charles to make a collection of plants growing on the property. He referred to this study in *On the Origin of Species.*

The local schoolteacher in Downe, John Mumford, was hired to teach writing to Willy and Annie. Annie's penchant for writing is reflected in the book title *Annie's Box*, which refers to the case in which she kept her writing supplies (R. Keynes 2001). This book is essential reading for anyone interested in Annie's life and the Darwin household. It is my favorite Darwin book, and this chapter draws heavily on its contents.

IMPRESSIVE WORK SCHEDULE

In the first years at Down House, Charles continued to read widely and make notes on his developing theory. He consulted with Joseph Dalton Hooker, a botanist who was to become his closest friend, confidant, and sounding board for Darwin's ideas on evolution (Freeman 1978). In July 1844 Charles enlarged his 1842 essay on his theory regarding species. He eventually shared some extracts from it with his friend Asa Gray, a botanist at Harvard University, in 1857 (F. Darwin 1909). Charles wrote a note to Emma requesting that she spend £400 to publish this manuscript in the event of his sudden death (Burkhardt et al. 1985–, 3: 43–44). Later that year, *Geological Observations on Volcanic Islands* was published (C. Darwin 1844).

Geological Observations on South America was published in 1846 (C. Darwin 1846). This was the third volume in the *Geology of the Voyage of the Beagle*. Also in 1846, Charles began what was to become a tedious, eight-year study of every living and fossil barnacle species known. He was to examine, dissect, and describe over 10,000 species. This gave him an appreciation for variation, adaption, and taxonomy (Stott 2003). The painstaking dissections were done in his study, while Charles perched on a stool with a swiveling seat, working near a window that provided natural light. The microscope he used was made to his specifications.

CHARLES'S FATHER'S DEATH

Two years into this major project, Charles's father died. Dr. Robert Waring Darwin was a respected physician who operated a large medical practice out of his home, the Mount, in Shrewsbury. He was a mountain of a man at 6 feet 2 inches tall, weighing over 300 pounds. Robert also had a formidable presence and personality. Charles was in awe of his father, whom he respected greatly. Charles's own health deteriorated as he watched his father's decline. He traveled to Shrewsbury for his father's eighty-second birthday on 30 May 1848, and he visited his father for two weeks in late October (Desmond and Moore 1991). Dr. Darwin died on 13 November 1848, with his daughter Susan in attendance. Charles was too miserable to attend the funeral. He avoided funerals all his life, unless it was absolutely impossible for him not to attend. After a brief visit to the Mount after his father's death, Charles retreated to Down House and the comfort of Emma.

A WEALTHY FAMILY

The Charles Darwins were well off financially. By 1845 their annual income was about £1,400, which would be roughly equivalent to $117,000 in today's terms. Upon the death of Dr. Darwin, Charles and his brother Ras each inherited one-fourth of their father's estate. The four Darwin sisters each received one-eighth, under the gender-biased inheritance system of the time. This put Charles's inheritance at around £51,000, or $4.25 million in today's money, according to the calculations of Loy and Loy (2010).

CHARLES'S ILLNESS AND THE WATER CURE

Charles's depression and chronic vomiting became so debilitating that, in desperation, he finally decided to take a water cure at Dr. James M. Gully's spa at Malvern Wells, Worcestershire, near the border with Wales. This was also to be a family vacation. Charles and Emma brought the children, Miss Thorley, and servants. Charles arranged accommodations in a nearby villa called the Lodge (R. Keynes 2001). Dr. Gully's reputation was widespread, and his patients included some notable trend-setters: Thomas and Jane Carlyle, Alfred Tennyson, Florence

Nightingale, Charles Dickens, and William Gladstone, the future prime minister.

The hydrotherapy consisted of being wrapped in wet sheets, deluged with cold water from an elevated reservoir, and vigorously scrubbed with a brush. The intention was to improve one's circulation. Darwin claimed he felt better because of these treatments, and he eventually installed a shower station in his garden at Down House, where he took his morning scrub. The treatments and vacation lasted from 10 March to 30 June 1849. Dr. Gully also insisted that Charles give up snuff and prescribed a diet devoid of tasty foods (Desmond and Moore 1991). Darwin did indeed feel restored after these treatments, perhaps because this enforced holiday relieved the stress his work on evolution induced, and he once again began taking long walks around the countryside. Eventually he returned to Down House to resume his barnacle work.

ANNIE'S ILLNESS

In November 1849, eight-year-old Annie came down with scarlet fever, as did her younger sisters Henrietta and Elizabeth (Figure 4.1). Annie's sisters recovered, but Annie remained fragile. In 1850 the Darwin children included Willy, Annie, Etty (age six), Georgy (age four), Bessy (age two), Franky (age one), and Emma's newly born eighth child, Leonard. In late June 1850, Annie complained of feeling ill. She felt poorly for weeks at a time and cried frequently for no obvious reason. Her lessons with the governess, Miss Thorley, became a trial for both pupil and teacher. Berry picking with her Wedgwood cousins in August, and a seaside family vacation in October at Ramsgate, failed to rally her (Healey 2001). Emma took her to London in November, and again in December, to consult with Dr. Sir Henry Holland, Charles's second cousin, who also

FIGURE 4.1 Daguerreotype of Anne Elizabeth Darwin in January 1849, at 7 years and 10 months old, by London photographer Claudet. Darwin Museum, Down House.

happened to be physician to Queen Victoria and Prince Albert. He suggested that Annie had inherited Charles's digestive problems, and there was nothing he could do to help (Desmond and Moore 1991).

Annie became more and more withdrawn. Since the most prominent physician in the country had nothing to offer, Charles decided to try the less-than-orthodox methods of Dr. Gully. Charles felt that his chronic problems had benefited from the cold-water cure he had taken on several occasions, and he resolved to take Annie there as a last resort, if it became necessary. In the meantime he treated her at home with dripping sheets and a spinal wash, and by packing her in damp towels, rubbing her, and making her sweat, all under Dr. Gully's specific directions.

Annie had her tenth birthday on 2 March 1851 at Down House. Shortly thereafter Charles, Emma, Annie, and the entire household came down with the flu. Annie was struggling, and the decision was made to take her to Malvern Hills. Emma was pregnant with soon-to-be-born Horace, so she remained at home with the rest of the children. Charles took Annie, her sister Henrietta, and the family's Scottish nurse, Jessie Brodie, who was Annie's favorite servant, to Dr. Gully's spa on 24 March (R. Keynes 2001). They would soon be joined by Miss Thorley. Four days after lodging the patient and her entourage at Montreal House in Malvern and being assured that Annie was in good hands, Charles left for London to visit Charles Lyell, Erasmus, and other family and friends. He returned to Down House on 31 March.

The family was notified on 15 April that Annie had taken a turn for the worse, and Charles was back in Malvern by 17 April. Emma arranged for her sister-in-law, Frances ("Fanny") Wedgwood, to meet Charles at Malvern, and for her aunt, Fanny Allen, to come to Down House to help with Emma's impending delivery. Annie had a fever and was vomiting. Both Brodie and Thorley were inconsolable, and Henrietta was despondent. When Charles saw Annie's pitiful condition he collapsed. It was a no-win situation for Charles. He wanted to be at home supporting Emma, but this was his precious daughter and she was suffering terribly. Annie went in and out of consciousness and delirium. Charles tried to feed her, but she could not keep anything down. She was exhausted and vomiting "bright green" bile, but she was still alive, and Charles's spirits rose and fell with Annie's condition. He wanted to believe that she would rally.

Charles wrote to Emma each day, and sometimes more than once, with hourly reports of Annie's progress and failings. Emma pored over the letters and answered them with more questions. The British postal service was extremely efficient, with overnight deliveries between Malvern and Down House, and Saturday mail deliveries. The exchange of letters between

Charles and Emma during this trying time, described in *Annie's Box*, constitutes one of the loveliest and most moving examples of the devoted marriage they shared (R. Keynes 2001). The author, Randal Keynes, is a great-great-grandson of Charles Darwin via Charles's son George. Keynes is also the grand-nephew of economist John Maynard Keynes, who wrote about Leonard Darwin (see chapter 10).

Henrietta, only seven years old, kissed Annie goodbye and was escorted to the home of Aunt Caroline (Charles's sister) and Uncle Josiah Wedgwood III (Emma's brother), where Fanny's children were also staying. Meanwhile, Annie seemed to rally, and Dr. Gully said she had turned the corner. Charles telegraphed Erasmus to get the good news by messenger to Emma. Annie was eating, her pulse rate had improved, and her fever was gone. Charles and Fanny stayed with Annie all night. Annie struggled when her bladder had to be catheterized. When Brodie washed her face and hands, Annie kissed her. When given a little brandy and water she replied, "I quite thank you." Dr. Gully continued to say she was getting better. The next morning (Easter Monday) her pulse was fluttering and her diarrhea was foreboding. Charles wrote to Emma, "I wish you could see her now, the perfection of gentleness, patience and gratitude." Annie thanked Fanny for a sip of tea and pronounced it "beautifully good." Charles reported the ups and downs to Emma, but neither knew exactly what to expect as hope gradually faded. Emma wrote to Charles, "I fear, after your letter today, there is but one account to expect tomorrow" (R. Keynes 2001).

ANNIE'S DEATH

Tuesday was supposed to be the turning point, according to Dr. Gully, but Annie continued to lose strength, and so did Charles. He vomited as Annie deteriorated. Now even Dr. Gully was losing hope. On Wednesday, 23 April 1851, at noon during a rainstorm, her breathing slackened and Annie lay lifeless. Brodie and Miss Thorley collapsed. Fanny tried to comfort them, to no avail. Charles kissed his daughter for the last time. He broke down again when he wrote to Emma that Annie went "to her final sleep... without a sigh." Dr. Gully pronounced the cause of death as "Bilious Fever with typhoid character" to a weeping Charles. This was a symptomatic description, not a disease. It was not typhoid fever as we know it today, but rather consumption, the old name for tuberculosis (R. Keynes 2001), a bacterial infection. It may have been triggered by the flu Annie caught in March (Healey 2001). Charles, however, feared that it was some sort of hereditary weakness.

Charles needed to be with Emma, his source of all comfort. He could not stand to deal with a funeral for his beloved Annie. Fanny said he should go to Emma, and that she (Fanny), with the help of her husband Hensleigh Wedgwood, who had just arrived, would arrange the funeral and burial at Malvern. Emma received no letter on Wednesday, and she understood what that meant. Before Thursday's mail arrived with Charles's letter, Emma had penned a note to Charles that she gave to the postman for delivery in Malvern: "You must remember that you are my prime treasure (and always have been). My only consolation is to have you safe home to weep together." On Thursday Charles left instructions for everyone with Fanny and returned to Down House that evening. He arrived at the door at six thirty. Charles and Emma held each other and wept. At Charles's direction, Annie's tombstone in the Malvern churchyard reads "A dear and good child" (Figure 4.2).

Eleven-year-old William and seven-year-old Henrietta were devastated by Annie's death. It was especially difficult for Etty, as she was the last child to see Annie alive. Brodie, now nearly 60 years old, took the loss of her charge very hard. She left the Darwin household in 1851 to return to Scotland, although she continued to visit Down House (Freeman 1978). Darwin

FIGURE 4.2 Headstone of Anne Elizabeth Darwin in the church cemetery at Malvern. Photograph by James Moore, modified from Desmond and Moore (1991).

arranged for her to receive an annual pension for the rest of her life. She had been 49 years old when she joined their household in 1842. Miss Thorley grieved in private at her mother's home for a few weeks and then returned to the Darwins.

IN MEMORIAM

A week after Annie's death, Charles wrote a loving memoir, intended solely for Emma and him. Desmond and Moore (1991) describe it as "the most beautiful—and certainly the most intensely emotional—piece he would ever write." The following, quoted in its entirety, is from Burkhardt et al. (1985–), 5: 540–542:

> Our poor child, Annie, was born in Gower St on March 2nd 1841 and expired at Malvern at Midday on the 23rd of April 1851. I write these few pages as I think in after years, if we live, the impressions now put down will recall more vividly her chief characteristics. From whatever point I look back at her, the main feature in her disposition which at once rises before me is her buoyant joyousness, tempered by two other characteristics, namely her sensitiveness, which might easily have been overlooked by a stranger, and her strong affection. Her joyousness and animal spirits radiated from her whole countenance and rendered every movement elastic and full of life and vigour. It was delightful & cheerful to behold her. Her dear face now rises before me, as she used sometimes to come running down stairs with a stolen pinch of snuff for me, her whole form radiant with the pleasure of giving pleasure. Even when playing with her cousins when her joyousness almost passed into boisterousness, a single glance of my eye, not of displeasure (for I thank God I hardly ever cast one on her), but of want of sympathy would for some minutes alter her whole countenance. This sensitiveness to the least blame, made her most easy to manage & very good; she hardly ever required to be found fault with, and was never punished in any way whatever. Her sensitiveness appeared extremely early in life; & showed itself in crying bitterly over any story at all melancholy, or on parting with Emma even for the short interval. Once when she was very young she exclaimed "Oh Mamma, what should we do, if you were to die?"
>
> The other point in her character, which made her joyousness & spirits so delightful, was her strong affection, which was of a most clinging, fondling nature. When quite a Baby, this showed itself in never being easy without touching Emma, when in bed with her, & quite lately she would, when poorly, fondle for any length of time one of Emma's arms. When very unwell, Emma

lying down beside her seemed to soothe her in a manner quite different from what it would have done to any of our other children. So again, she would at almost any time spend half-an-hour in arranging my hair, "making it" as she called it "beautiful," or in smoothing, the poor dear darling, my collar or cuffs, in short in fondling me. She liked being kissed; indeed every expression in her countenance beamed with affection & kindness, & all her habits were influenced by her loving disposition.

Besides her joyousness thus tempered, she was in her manners remarkably cordial, frank, open, straightforward, natural and without any shade of reserve. Her whole mind was pure & transparent. One felt one knew her thoroughly and could trust her: I always thought, that come what might, we should have had in our old age, at least one loving soul, which nothing could have changed. She was generous, handsome & unsuspicious in all her conduct; free from envy and jealousy; good-tempered and never passionate. Hence she was very popular in the whole household, and strangers liked her & soon appreciated her. The very manner in which she shook hands with acquaintances showed her cordiality.

Her figure & appearance were clearly influenced by her character: her eyes sparkled brightly; she often smiled; her step was elastic & firm; she held herself upright, & often threw her head a little backwards, as if she defied the world in her joyousness. For her age she was very tall, not thin, & strong. Her hair was a nice brown & long; her complexion slightly brown; eyes dark grey; her teeth large & white. The daguerreotype is very like her, but fails entirely in expression: having been made two years since, her face had become lengthened & better looking. All her movements were vigorous, active & usually graceful; when going round the sand-walk with me, although I walked fast, yet she often used to go before pirouetting in the most elegant way, her dear face bright all the time, with the sweetest smiles.

Occasionally she had a pretty coquettish manner towards me, the memory of which is charming: she often used exaggerated language, & when I quizzed her by exaggerating what she had said, how clearly can I now see the little toss of the head & exclamation of "Oh Papa, what a shame of you."—She had a truly feminine interest in dress, & was always neat: such undisguised satisfaction, escaping somehow all tinge of conceit & vanity, beamed from her face, when she had got hold of some ribbon or gay handkerchief of her Mamma's.—One day she dressed herself up in a silk gown, cap, shawl & gloves of Emma, appearing in figure like a little old woman, but with her heightened colors, sparkling eyes & bridled smiles, she looked, as I thought, quite charming.

She cordially admired the younger children; how often have I heard her emphatically declare "What a little duck Betty is, is not she?"

She was very handy, doing everything neatly with her hands: she learnt music readily, & I am sure from watching her countenance, when listening to others playing, that she had a strong taste for it. She had some turn for drawing, & could copy faces very nicely. She danced well, & was extremely fond of it. She liked reading, but evinced no particular line of taste. She had one singular habit, which, I presume, would ultimately have turned into some pursuit; namely a strong pleasure in looking out words or names in dictionaries, directories, gazetteers, & in this latter case finding out the places in the Map: so also she would take a strange interest in comparing word by word two editions of the same book; and again she would spend hours in comparing the colours of any objects with a book of mine, in which all colours are arranged & named.

Her health failed in a slight degree for about nine months before her last illness; but it only occasionally gave her a day of discomfort: at such times, she was never in the least degree cross, peevish or impatient; & it was wonderful to see, as the discomfort passed, how quickly her elastic spirits brought back her joyousness & happiness. In the last short illness, her conduct in simple truth was angelic; she never once complained; never became fretful; was ever considerate of others; & was thankful in the most gentle, pathetic manner for everything done for her. When so exhausted that she could hardly speak, she praised everything that was given her, & said some tea "was beautifully good." When I gave her some water, she said "I quite thank you"; and these, I believe were the last precious words ever addressed by her dear lips to me.

But looking back, always the spirit of joyousness rises before me as her emblem and characteristic: she seemed formed to live a life of happiness: her spirits were always held in check by her sensitiveness lest she should displease those she loved, & her tender love was never weary of displaying itself by fondling & all the other little acts of affection.

We have lost the joy of the household, and the solace of our old age:—she must have known how we loved her; oh that she could now know how deeply, how tenderly we do still & shall ever love her dear joyous face. Blessings on her.

April 30, 1851

VISIT TO ANNIE'S GRAVE

In August 1863, 12 years after Annie's death, Charles was very unwell, and Emma insisted that they journey to Malvern for treatments and to see Annie's grave. The grave had been relocated and could not be found among the overgrown shrubs. After getting precise directions from William Darwin Fox and the owner of Montreal House, where Annie died, Emma, with Charles in tow, found the gravestone (R. Keynes 2001).

HOOKER'S LOSS

While taking water treatments at Malvern, Charles received a letter from his closest friend, Joseph Dalton Hooker, stating that

Hooker's six-year-old daughter, Maria, had died. The letter from 28 September 1863 showed the closeness of their relationship, as Hooker recalled Charles's pain over the loss of Annie. "Dear dear friend, My darling little 2[d]. girl died here an hour ago, & I think of you more in my grief, than any other friend" (Burkhardt et al. 1985–, 11: 640). Hooker acknowledged receipt of Darwin's kind note on 1 October, but Darwin's letter has been lost.

CHARLES'S DISBELIEF COMPLETE

The death of his father in 1848, and especially Annie's death in 1851, alienated Charles completely from any shred of religious belief (Moore 1989; Desmond and Moore 1991). Intellectually, he had already rejected the idea of "revelation." Charles admitted that Christianity was "not supported by evidence" and further stated that "I never gave up Christianity until I was forty years of age." He was 42 when Annie passed away and Annie's death was simply the coup de grâce for a preexisting rejection of faith. Charles wrote in his *Autobiography* that "disbelief crept over me at a very slow rate, but was at last complete. The rate was so slow that I felt no distress, and have never since doubted even for a single second that my conclusion was correct." Charles felt it was irrational to envision a just and merciful God that would allow such suffering to befall innocent children, or who would condemn to Hell nonbelievers such as his father, his brother, and many of his friends, all of whom he recognized as good people (Moore 1989). Charles described himself as an agnostic, but he would never attack another person's belief. After all, he was emotionally bound to his devout wife. Emma took what comfort she could from her religious belief, with its promise of an afterlife. Charles had no such delusions.

BARNACLES

Once again, as he had done after his father's death, Charles sought refuge in his barnacle work. In June 1851 his *Monograph of the Fossil Lepadidae* was published (C. Darwin 1851a). His *Monograph of [Recent] Lepadidae* (C. Darwin 1851b) appeared the next year, but it also bore an 1851 publication date. The remaining volumes, *Monograph of the Fossil Balanidae* and *Monograph of [Recent] Balanidae* (C. Darwin 1854a, 1854 b followed in 1854 (Freeman 1977).

GONE BUT NOT FORGOTTEN

Twenty-five years after Annie's death, Charles wrote in his autobiography that "tears still sometimes come into my eyes, when I think of her sweet ways."

MARY ELEANOR DARWIN (1842–1842)

Sadly, this will be a very short chapter, because the third child of Charles and Emma Darwin lived only 23 days, from 23 September until 16 October 1842.

MOVE TO DOWN HOUSE

The growing family and its several servants, the crowded and polluted city, and Charles's general ill health led to the decision to find a house in the country. Charles wanted a place far enough removed from the city so that he could avoid unnecessary socializing, but close enough so that he could get to London and back in one day. Charles and Emma first saw Down House on 22 July 1842. Located 16 miles south of London, it was a two-hour train and carriage ride from London Bridge Station. It was situated among 18 acres on Luxted Road, Downe, Orpington, Kent. Beginning in 1842 the village of Downe was spelled with an *e*, but the Darwins did not add an *e* to Down House. The population of Downe was 444 in the 1841 census (Freeman 1978). With help from Charles's father they purchased the property for £2,020. A pregnant Emma, children Willy and Annie, and nursemaid Elizabeth "Bessy" Harding moved in on 14 September, so Emma could get settled before giving birth; Charles followed three days later (Browne 1995). Emma and Charles each had their own bedroom, probably so that when

Charles felt unwell he would not disturb Emma's rest. There was no indoor bathroom (R. Keynes 2001).

Charles employed a butler, a footman, and two gardeners, while Emma had a cook, a kitchen maid, a laundry maid, a housemaid, a nurse, and at least one nursemaid. The Darwins treated their servants very kindly, which elicited a great deal of devotion and loyalty to the family, especially the children. Most of the maids were local girls from the nearby villages (R. Keynes 2001).

MARY'S BRIEF LIFE

Emma moved into Down House on 14 September, and Charles joined her a few days later. The Downe village surgeon delivered the small and feeble baby. Emma reckoned that she was in her thirty-sixth week of pregnancy when she delivered (R. Keynes 2001). A normal pregnancy usually lasts 40 weeks. Emma recovered rapidly in her new country surroundings. Baby Mary was baptized in the village church at nine days of age. She was buried in the churchyard on 19 October (Desmond and Moore 1991). The parents grieved, but they took solace in the fact that Mary did not seem to have suffered during her very short life. The cause of death is not known. The first few months or so were a rather depressing beginning to the Darwins' new life at Down House. Eventually, though, they became more absorbed in making changes to the house and gardens. Emma became pregnant with Henrietta in January 1843. The premature death of Mary left a gap of two and a half years between the ages of Annie and Henrietta.

HENRIETTA EMMA DARWIN

(1843–1927)

The Darwins' fourth child was born on 25 September 1843 at Down House. Emma nursed her new baby, as she had done with the other children, and Charles affectionately referred to Emma as "my dear old Titty" (Desmond and Moore 1991). He wrote that he felt "very much in love with my own dear three chickens." Henrietta was called by a variety of nicknames, but "Etty" seemed to stick. Emma was devoutly Unitarian, the religion of the Wedgwood family. She took her children to church, read the Bible to them, and had the children baptized and confirmed in the Church of England.

In January 1844, while Emma was accommodating a new infant into her life, Charles wrote to his fairly recent botanist friend Joseph Dalton Hooker, who had just returned from a four-year voyage of discovery to the Antarctic. Darwin confided his view that he thought species could change, and this felt to him like "confessing a murder" (Burkhardt et. al. 1985–, 3: 2). Charles's intellectual work was the constant soundtrack playing in the background of his children's lives.

A SICKLY CHILDHOOD

Etty remembered her father as a delightful playmate, and she also enjoyed trying on her mother's skirts and jewelry, especially when playing with her Wedgwood cousins. In late 1849, six-year-old Etty came down with scarlet fever, along with Annie and their younger sister Elizabeth. Etty earned a reputation as a rather sickly child, but she remembered the happy times of playing in the streams and fields behind Malvern while her father took the cure (Browne 1995). She also recalled listening outside her father's shower house in the garden, hearing his groans as the cold water rained down on him during his self-administered water cure at Down House.

In October 1850 Charles bought Annie and Etty a canary (R. Keynes 2001). He seemed to enjoy studying its acquisition of song as much as the children enjoyed playing with their pet. Although seven-year-old Etty was at Malvern with Annie before Annie died, Etty was never quite sure what was happening. Annie's death in 1851 was a very strong emotional blow to Etty. She had always considered Annie as the good girl with the pleasant disposition. Etty thought of herself as naughty and temperamental and worried that she would not make it to heaven to see Annie (Browne 1995). Later that year, as a happier distraction, Charles and Emma brought Etty and her brother George to the Great Exhibition in London (Figure 6.1). The family stayed with Charles's brother Erasmus.

After Charles finished his barnacle tome in 1854, he turned his attention to breeding fancy pigeons. The artificial selection that breeders used to create bizarre forms of the common pigeon became a proxy for the natural selection that Darwin had already observed acting in nature. Henrietta took great delight in having a pigeon coop in the garden. She helped her father with breeding experiments and learned the temperament and personality of various breeds. She became very upset, however, when, without Etty being consulted, her pet cat was "executed" for the crime of murdering pigeons (Montgomerie 2009).

Emma had a rather relaxed view of education for her daughters. Henrietta was a very sickly child, but she could have

FIGURE 6.1 *Left*: Daguerreotype of nearly eight-year-old Henrietta Emma Darwin, taken in August 1851, by Richard Beard in London. Henrietta's expression and dark dress may reflect mourning for the recent death of Annie. *Right*: Henrietta Emma Darwin, somewhat older. Sophie Gurney.

benefited from more rigorous schooling. Nonetheless, the traditional Wedgwood method of reading aloud to the girls and encouraging them to use the library was successful in Henrietta's case. She eventually developed into an excellent editor of her father's writings, despite lacking a university education (Healey 2001).

Henrietta had serious dental problems, as did several other members of the Darwin household (Loy and Loy 2010). Perhaps this could be attributed to the rather rich foods they were served, such as sauces and sweets. In 1854, when Henrietta was ten years old (Figure 6.1), she was taken for an appointment with Darwin's dentist in London, James Robinson, who also happened to be Prince Albert's dentist. Robinson extracted four of Henrietta's teeth. Four years later another London dentist crafted a gold plate to realign her teeth.

ILL HEATH AS A LIFESTYLE

When Henrietta was 13 years old she developed a cold and a fever, and the doctor recommended that she should have breakfast in bed for a while. In *Period Piece*, Gwen Raverat (1952) wrote that Etty "never got up to breakfast again in all her life." Chapter 7 of that lovely book is devoted to Aunt Etty and her valetudinarian habits. Raverat remembered Aunt Etty as a "lady," that is "as [one of the] people who did not do things themselves." As an example, Raverat revealed that 86-year-old Henrietta confided to her that she had never made a pot of tea in her life. Henrietta had nothing to do, and, Raverat wrote, "ill health became her profession and absorbing interest."

Etty's eccentricities included asking her long-suffering maid, Janet, to count Etty's leftover prune pits and to cover her left foot with a silk scarf, "because it was that amount colder than her right foot." When colds were making the rounds, Etty wore a kitchen strainer stuffed with eucalyptus-soaked cotton on her face, like a snout (Figure 6.2). She seemed oblivious to the fact that visitors struggled to stifle their laughter. "We all laughed at her and we all adored her," wrote Raverat. "She enjoyed her profession [of invalid] very much."

In March 1857, while Charles was struggling to write the enormous book that he was tentatively calling "Natural Selection" and dealing with his own ill health, Emma took a sickly Henrietta to Hastings, Sussex. She thought that the sea air would help Etty feel better and allow Charles to not be distracted in his writing. After two months, Henrietta was no better. Emma took her to Dr. Edward Lane's hydropathic establishment at Moor Park, which was much closer than Dr. Gully's Malvern facility (Desmond and Moore 1991). After two weeks Charles relieved Emma, both to watch over Etty and to take some treatments himself. He played backgammon with his daughter, but his concern for her was "too agitating." Charles improved more than Henrietta, and both invalids were soon back at Down House. Charles even took on new responsibilities, becoming a justice of the peace in Downe. In November 1857 Charles again took himself to Moor Park for a week's rest.

Things went from bad to worse in 1858. Henrietta contracted diphtheria, and Dr. Lane was on trial for adultery, accused by a patient. Charles didn't believe the charges against his favorite therapist (Desmond and Moore 1991). Dr. Lane was eventually acquitted and moved his practice to Sudbrook Park, Richmond.

The arrival of a letter from Alfred Russel Wallace on 18 June 1858 introduced further turmoil into Charles's life (see chapter 12). On 18 July, Charles's oldest sister Marianne (Mrs. Henry Parker), age 60, died. As usual, Charles did not attend the funeral. Her adult children (four sons and one daughter) were adopted by Charles's unmarried sisters Emily Catherine ("Catty") and Susan and went to live at the Mount in Shrewsbury (Loy and Loy 2010). All of this worry took its usual toll on Charles's health.

In 1859, 16-year-old Henrietta complained to her father about the manner in which Madame Grunt, a Swiss teacher of French and German, scolded the children, especially Henrietta herself. Charles fired the teacher immediately (Loy and Loy 2010). In 1860 the Darwin family took an extended summer vacation to visit Emma's sisters in Hartfield, Sussex. This was mostly for Henrietta's benefit, since she remained chronically ill, probably as a result of a typhoid-like fever. Charles spent much of the time there studying the local plants, including the insectivorous sundews (*Drosera*), which became the subject of his book *Insectivorous Plants* 16 years later.

Henrietta 's ill health, manifested by fever and chronic vomiting, continued off and on through 1860. She received dozens of doses of calomel (mercury chloride) and grey powder (mercury

FIGURE 6.2 *Left*: Henrietta Darwin wearing her cold mask: a kitchen strainer stuffed with eucalyptus-soaked cotton, designed to kill cold germs. Drawn by Henrietta's niece, Gwen Raverat, from *Period Piece* (Raverat 1952), *Right*: Henrietta Darwin Litchfield, holding a thermometer, covered her long-suffering and patient husband Richard with a cloth when airing the room, so he would not catch a cold. Drawn by Henrietta's niece, Gwen Raverat, from *Period Piece* (Raverat 1952). Both illustrations used with permission from Faber and Faber, Ltd. and W. W. Norton.

and chalk) in less than half a year (Loy and Loy 2010). Her suffering upset Charles, who was too ill to visit Professor Henslow, the man who influenced his career more than any other, before Henslow died on 18 May 1861. Charles, continuing his usual practice, did not attend his beloved professor's funeral. On some bad days, Henrietta vomited 19 or more times. She continued to be a sickly invalid at age 18 and required attendance around the clock. Fevers and digestive problems racked her frail body. This caused Charles and Emma much anxiety. They took comfort in each other, and Charles became very nervous whenever separated from Emma.

Etty and Charles continued to have stomach problems, and several therapeutic holidays were taken, with little effect. Charles and Emma feared that Etty would die. Dr. Henry Holland told them to prepare for the worst, but the worst did not come. Etty continued to receive lavish attention from her mother and father, living as a semi-invalid and a constant source of worry for Charles and Emma. When Etty was sick, Charles would play backgammon with her, and Emma read aloud (Browne 2002). Being nursed by Emma was a very comfortable situation, and it may have had some impact on a subconscious desire to remain unwell, a situation that may have applied to Charles in addition to Henrietta. He continued to worry that his digestive problems were inherited by Henrietta and all of his subsequent children (Colp 2008), and that this weakness may have been a consequence of his marriage to his cousin (R. Keynes 2001).

Darwin's granddaughter Gwen Raverat remarked that "at Down, ill health was considered normal." She then elaborated: "The trouble was that in my grandparents' house it was a distinction and a mournful pleasure to be ill. This was partly because my grandfather was always ill, and his children adored him and were inclined to imitate him; and partly because it was so delightful to be pitied and nursed by my grandmother." Raverat further pointed out that there were hundreds of letters written by Emma and Aunt Etty, "and every one of them also contains dangerously sympathetic references to the ill health of one, or several, of the family."

GRADUAL IMPROVEMENT

Charles took Etty fishing in the summer of 1861, near the village of Torquay on the Devon coast (Desmond and Moore 1991). The sea air seemed to help her. By August 1861 Henrietta began a slow climb to better health, but with some setbacks. Charles utilized the holiday to spend time on his belly, observing insects pollinating wild orchids. At home, Etty often followed him into the hothouse at Down and seemed to enjoy the

FIGURE 6.3 *Left*: Nineteen-year-old Henrietta Emma Darwin, photographed by S. J. Wiseman, Southampton, about 1862. Cambridge University Library. *Right*: Henrietta Emma Darwin. Cambridge University Library.

orchids, sundews, and other wonders he studied. When Etty was 18 years old she helped her father correct the proofs for his orchid book (C. Darwin 1862) (Figure 6.3).

HER FATHER'S SECRETARY

Meanwhile, Charles continued to experiment with *Drosera rotundifolia*, the common sundew, and its digestive process. When Charles's own digestion was playing up, Henrietta and Emma handled his letter writing (including forging his signature) and recorded his notes by dictation. Henrietta assisted her father in correcting the proofs for his largest book, *The Variation of Animals and Plants under Domestication*, published in January 1868. For this service she was paid £20 (Browne 2002).

At 27 years old, Henrietta was an eligible young woman and, traveling with some relatives, visited Cannes in 1870 to have a look around. When she returned home she began calling herself "Harriot," which Emma soon put a stop to by referring to her as "Body." Charles sometimes called her "Hen." Etty tried to shift Charles from "Papa" to "F" (for "Father"), to which he replied, "I would as soon be called 'Dog.' "

Henrietta helped Charles edit his manuscripts, especially *The Descent of Man*. She was no mere proofreader, as her father charged her with significant responsibilities. In a letter to Henrietta on 8 February 1870, Charles wrote: "My Dear H. Please read the Ch. first right through without a pencil in hand, that you may judge of general scheme; as, also, I particularly wish to know whether parts are extra tedious; but remember that M.S. is always much more tedious than print.... After reading once right through, the more time you can give up for deep criticism or correction of style, the more grateful I shall be.—Please make any long corrections on separate slips of paper, leaving narrow blank edge, & pin them to margin of each sheet, so that I can turn each back, & read whilst still attached to its proper page.—This will save me a world of troubles.... You are a very good girl indeed to undertake the job" (Burkhardt et al. 1985–, 18: 25).

Henrietta was the guardian of propriety and often softened Charles's prose to tone down potential controversy. She quarreled with her father over punctuation and was a good editor. He grew to respect and value her advice. Even Thomas Henry Huxley was impressed with Miss Etty's eagle-eyed editorial criticism when she spotted some errors in Huxley's work. *The Descent of Man* was a critical, popular, and financial success, and Charles offered Etty a monetary reward (£30) from his royalties as recompense for her efforts (Burkhardt et al. 1985–, 19: 199; Desmond and Moore 1991). The first edition appeared in 1871, followed by a second edition in 1874. The book also spawned many personal attacks on Charles, and he was depicted in cartoons as an ape (Figure 6.4). Such caricatures reinforced

FIGURE 6.4 Charles Darwin as an ape. Cartoon from the *London Sketchbook*, 1874.

the connection in the public's mind between Darwin and the theory he proposed (Browne 2003).

On the grounds at Down House, prim and proper Henrietta waged war against the stinkhorn fungus, *Phallus impudicus*, which had the misfortune to resemble its generic namesake. The offending phalli were gathered and burned in the drawing room fireplace for the moral protection of the housemaids (Raverat 1952).

A MATE FOR HENRIETTA

No one in the family expected sickly, eccentric Henrietta to find a husband, but the heart has its own agenda. Henrietta astonished her family when she announced her engagement in

June 1871. Richard Buckley Litchfield was a scholar and philanthropist who worked on the legal end of the Ecclesiastical Commission, which managed Church of England property. He founded the Working Men's College of London (which eventually became Birkbeck College), where he taught music, mathematics, and science in his spare time. He was 12 years older than Henrietta, and heavy set, with a long, thick, brown beard and a sweet smile (Desmond and Moore 1991). Henrietta, who had taken singing and piano lessons as a child, was attracted to his musical side (he loved singing and conducting). She was also aware that beards were a secondary sexual characteristic, information gleaned from reviewing the parts about sexual selection in *The Descent of Man*. Henrietta and Richard first met at the London home of Judge Vernon Lushington around 1869 (Freeman, 1978).

Young and in love, Henrietta recorded her most intimate thoughts in her diary, which has only recently become available to the public (Burkhardt et al. 1985–, 19: appendix 6, 801–807). On 4 July 1871 she mused: "To make him happy. To guard my health for him. Not to forget that I am a social animal because I have my own little life to live—to cultivate my mind—& last to try & make up to my people for losing me—to show the gratitude I do feel—& to feel it more. I have so often been a wretch & they have loved me so." The entry for 9 July reveals Henrietta's rather unexpectedly passionate side: "Is it love when I think about him day & night—when I wonder what he thinks on every conceivable subject—when I feel my day made bright & happy by one short letter. I want him to take me in his arms & say I shall never leave him—I long for him to strike the match which is to kindle me.... Oh if he had known under my icy manner how I longed for him just to take my hand just for one moment."

Brother George was Henrietta's closest friend, and she wrote the following to him: "Please prepare your mind for the most tremendous piece of news concerning myself I could tell you.

The supreme crisis of my life. I am going to be married to Mr. Litchfield. You will say that I don't know him,—that was true a fortnight ago when he asked me, but since then I do, & he has made me believe that he does care for me, as I have dreamt of being loved, but never expected that supreme happiness to fall to my lot" (Healey 2001).

Richard wrote to Vernon Lushington, his fellow teacher at the Working Men's College, that he was going to marry Henrietta Darwin, whom he not so romantically described as "very wise & good, not particularly beautiful, perhaps not beautiful at all." A male acquaintance of Henrietta's and Richard's considered Henrietta's face to be a "feminine and tender" reflection of her father's (Loy and Loy 2010).

The Darwins depended on Erasmus, who lived in London, to assure them that Richard was suitable for Henrietta. Litchfield was impoverished—by Darwin standards—but he earned £1,000 annually, and he was not a gold digger (Loy and Loy 2010). As her father had done with Emma, Henrietta told Richard that she did not accept the concept of a personal God. That seemed to suit Richard just fine. After a two-month courtship, they were married in a simple ceremony at the church in Downe on 31 August 1871. No friends or relatives were invited, in order to provide a calm setting for Charles. A few of Litchfield's friends found out about the wedding, however, and appeared in the church to surprise him. Darwin's ever-vigilant butler Parslow was astonished by this. Charles gave away the bride and William and Leonard served as witnesses (Loy and Loy 2010). The Litchfields honeymooned in Europe and, of course, both became ill. Henrietta relapsed many times in the next several years.

POLLY THE DOG

When Henrietta married and moved to London with her new husband, her fox terrier, Polly, adopted Charles (Figure 6.5).

FIGURE 6.5 Henrietta Emma Darwin in a window at Down House, with her fox terrier Polly. The dog adopted Charles Darwin after Henrietta married Richard Litchfield in 1871 and moved to London.

Polly's basket was near the fire in Charles's study, and she would follow him around the Sandwalk daily (Townshend 2009). When Charles returned after an absence, Polly would go berserk with excitement at his reappearance. Charles was thinking about Polly when, in *The Descent of Man* (1871), he equated "the feeling of religious devotion" to the "deep love of a dog for his master" or "of a monkey to his beloved keeper." Polly was the first animal to appear on the pages of *The Expression of Emotions in Man and Animals*. She is seen eyeing a cat on an out-of-view table, with her paw lifted, ready to pounce. Polly was old and ill at the time of Darwin's death and had to be put down by Francis a few days later. Polly is buried under an apple tree at Down House (Townshend 2009).

MARRIED LIFE

At age 62, Charles was not entirely thrilled with Henrietta's departure, since there was now one less female to tend him. Charles wrote to Henrietta and offered the following advice, preserved in *Emma Darwin: A Century of Family Letters, 1792–1896*, volume 2, edited by Henrietta Litchfield: "Keep her [Emma] as an example before your eyes, and then Litchfield will in future years worship and not only love you, as I have worship[ed] our dear old mother. Farewell, my dear Etty. I shall not look at you as a really married woman until you are in your own house. It is the furniture which does the job. Farewell, Your affectionate Father, Charles Darwin." The letter is dated 4 September 1871 from Down House. Etty was the only Darwin daughter to marry (Berra et al. 2010a, 2010b). Charles continued paying Henrietta an allowance of £350 annually after her marriage, and he also provided the Litchfields with £5,000 worth of stocks as a wedding gift (Loy and Loy 2010).

Probably because of her own hypochondria, Henrietta decided that Richard, too, was delicate, and he found it easier to play along than to resist. When Henrietta aired a room, Richard would be covered with a dust cloth to prevent a draft from reaching him (Figure 6.2). When the danger was over, he would be uncovered (Raverat 1952).

Henrietta continued to help with her father's work. She corrected and edited the proofs for *The Expression of Emotions in Man and Animals*, which was published in November 1872. Even Richard Litchfield contributed to Darwin's research in a way that must have pleased Henrietta. He and his father-in-law discussed the origin of music and singing as a courtship ritual (Loy and Loy 2010). Henrietta also helped with a rewrite of the second edition (1874) of Darwin's *Coral Reefs* book.

In the summer of 1873, Charles was working on the insectivorous *Drosera*. Thomas Henry Huxley, on the other hand, was overworked, underpaid, exhausted, depressed, and nearly broke. Charles, Emma, Ras, and 18 scientific friends contributed money to send Huxley on a vacation. Huxley's family of seven children visited the Darwins, and Emma looked after them while Huxley, accompanied by Joseph Dalton Hooker, departed for mainland Europe and a hike through the Alps. This kind of help goes far beyond professional respect. The personal devotion and loyalty, even love, among such prominent scientific colleagues is amazing and almost unimaginable today. Huxley's gratitude was effusive (Desmond 1997).

Also in the summer of 1873, Richard Litchfield brought Henrietta, along with his singing class of 70 young workers, to Down House. A large party ensued, with singing, dancing, games, and food. A good time was had by all, even Charles (Browne 2002). Henrietta and Richard also attended a party in 1873 at the London home of writer George Eliot (the pseudonym of Mary Ann Evans), whom they both admired. She, somewhat shockingly for the time, lived openly with her lover, George Henry Lewes. Charles himself paved the way for the invitation to the Litchfields.

THE ANTIVIVISECTIONIST MOVEMENT

In 1875 Henrietta became enamored with the antivivisectionist movement and asked her father to sign an antivivisectionist petition. In fact, Charles and Emma had written a pamphlet in 1863 protesting the use of steel vermin traps, because of the devices' inhumane, toothed-studded jaws (Loy and Loy 2010). Darwin nonetheless refused to sign the petition and warned Henrietta that attempts to interfere with animal experiments would have negative effects on physiology (Desmond and Moore 1991). Darwin and Huxley, neither of whom did live-animal experiments, realized that regulation was superior to restriction. Hooker, as president of the Royal Society, supported their position. They swayed Henrietta and her lawyer husband to their side. Richard Litchfield even drafted a blocking bill to the antivivisectionist bill. Darwin testified before the Vivisection Commission in November 1875. Huxley was not pleased with the outcome, making the comparison that a boy could torture a fish and a frog by angling with live frogs as bait, but a teacher demonstrating blood circulation in the web of a frog's foot could be punished (L. Huxley 1900). Similar complaints are heard today by scientists who must jump through animal ethics committees' hoops in order to work with live animals.

Henrietta's feelings about animal cruelty became confounded with her strident anti-Catholicism, in a story related by Raverat:

> AUNT ETTY. It's about a priest who rides so fast to give a man absolution before he dies, that he kills his horse under him. Isn't it horrible!
>
> GWEN RAVERAT. Well, I suppose they believe that absolution matters more than anything else.
>
> AUNT ETTY. But doesn't the Horse matter? Doesn't Cruelty matter? How can they think...

Raverat also reported an episode in Cannes, where Henrietta observed some thugs kicking an injured dog. She intimidated both the unruly mob and a policeman into submitting to her will. Like her father, Henrietta could not reconcile the problem of suffering with the concept of a personal God (Browne 2002).

HENRIETTA THE EDITOR

In the summer of 1879, Darwin toiled for six weeks on an essay about the life of his famous physician grandfather. Henrietta, ever the guardian of the family name, felt that the manuscript was too long, too honest, and contained too much wine, women, illegitimacy, and irreligiosity. She wanted to protect her mother from any embarrassment and, especially, from Erasmus Darwin's comment that "Unitarianism is a featherbed to catch a falling Christian" (Moore 1989; Desmond and Moore 1991). Henrietta marked up the manuscript to show her father what to delete, and papa complied (Browne 2002). *Erasmus Darwin* (C. Darwin 1879) was published in the fall. The publisher, John Murray, was pleased with the final result, but Charles vowed never again to be diverted from his proper scientific work. Fara (2012) explored the science and poetry of Charles Darwin's polymath grandfather and demonstrated that Erasmus Darwin was a significant historical figure long before his grandson wrote about many of the same ideas (Stott 2012).

Henrietta provided a similar service while Charles was writing his *Autobiography*. She edited the text in red pencil, deleting references to financial matters and to living persons who might be offended. Much of this familial censorship was for Emma's benefit. Later in life Henrietta reflected that because her mother was socially disinclined and her father was ill so often, her upbringing was rather restricted. This led to her self-described awkwardness, shyness, and feelings of being an outsider. She described Emma as "serene but somewhat grave" and noted that "the jokes and the merriment would all come from my father" (H. Litchfield 1915, 2: 45).

DEATH OF UNCLE RAS

Uncle Ras, Charles's bachelor brother, was a proponent of women's education, and Henrietta was attracted to his intellectual circle of friends. Henrietta and Richard Litchfield lived near Erasmus and, once Henrietta had established her household, Charles and Emma often stayed with them instead of at Erasmus's house. This was to keep an eye on Henrietta's health, which improved somewhat after her marriage. The Litchfields were also frequent visitors at Down House. Henrietta, and Fanny Wedgwood and her daughter, stayed with Erasmus to comfort him at the end of his life. He died quietly on 26 August 1881. There is speculation that Ras was in love with Fanny Wedgwood (Desmond and Moore 1991; Browne 2002; Loy and Loy 2010), but this is fodder for another book. Charles did attend his brother's funeral on 1 September 1881, when Ras was laid to rest in St. Mary the Virgin's churchyard at Downe (Desmond and Moore 1991).

DEATH OF FATHER

Charles's heart disease caught up with him on 15 December 1881, when he was laid low by chest pain. On 7 March 1882 he had more cardiac distress, and Dr. Sir Andrew Clark attended him on 10 March. Dr. Sir Norman Moore came to Down House on 19 March. Charles rallied for a few days, but he had more chest pain on 4–6 April and again on 11–12 April. The Litchfields arrived for dinner on 15 April; Charles passed out briefly and was revived with brandy. He felt well enough to take an assisted walk on 17 April. But the next evening he had more chest pain, and he took amyl nitrate capsules (a vasodilator) (Desmond and Moore 1991). Emma and Bessy were at home with him.

Charles knew the end was near. He whispered to Emma, "My love, my precious love, tell all my children to remember how good they have always been to me." He further told Emma, "I am not the least afraid to die" (Browne 2002). Dr. Charles Henry Allfrey arrived at 2:00 a.m. on 19 April, applied mustard plasters to Charles's chest, and departed after Charles took some breakfast about 8:00 a.m. After vomiting, Charles gasped,

"If I could but die." Frank arrived by 10:00 a.m. and Henrietta joined the family gathering at 1:00 p.m. Charles was able to recognize his children. Frank offered brandy and Henrietta rubbed his chest. Charles remarked that "you two dears are the best of nurses." Emma, who had finally agreed to rest, was called and held his head on her chest. Charles passed away at 4:00 p.m. on 19 April 1882, surrounded by Emma, Frank, Bessy, and Henrietta. Details of his funeral and burial in Westminster Abbey on 26 April 1882 are given in chapter 3.

ANOTHER BRUSH WITH DEATH

After Darwin's death, life eventually returned to its new normal for the Darwin siblings. Henrietta sporadically continued to be unwell. A bizarre poisoning event occurred in 1891, when Henrietta accidentally swallowed a small quantity of liniment that she had mistaken for her medicine. She nearly died, but a strong emetic saved the day. Hot water bottles applied to her feet, which were numbed by the poison, burned her skin and her recovery took weeks (Loy and Loy 2010).

DEATH OF HUSBAND

Richard Litchfield had other interests beyond his legal work and music. He authored a biography of Thomas Wedgwood, the chemist, who was arguably the inventor of photography (R. Litchfield 1903; Prodger 2009; Berra et al. 2010b). When Richard died in 1903, Henrietta moved to Burrow's Hill, Gomshall (in Surrey), and lived a relatively quiet life, interacting with her Darwin and Wedgwood relatives.

The Litchfields produced no children. Berra et al. (2010b) suggested that this unexplained infertility may have been a reflection of some genetic problem in the consanguineous Darwin-Wedgwood lineage. It may also have resulted from the megadoses of mercury Henrietta received as a child and a young woman, since high blood levels of mercury are known to be associated with male and female infertility (Choy et. al. 2002). We may never know for certain, though.

Henrietta wrote a privately printed biography of her husband (H. Litchfield 1910). She was especially influential in the lives of her favorite brother George's daughters, Gwen and Margaret, who became highly successful in their own careers (Healey 2001).

HENRIETTA'S LEGACY

She edited a collection of family letters that became a tribute to and biography of her mother, who died on 2 October 1896 at the Grove, Cambridge. The two-volume collection, entitled *Emma Darwin: A Century of Family Letters 1792–1896*, was privately printed in 1904 and subsequently published in 1915 (H. Litchfield 1915). In a letter to her brother Horace, who had sent some of their mother's letters to Henrietta for inclusion in Emma's biography, she wrote: "The object of the book is mainly to leave a record of an exceptionally beautiful character & of course how she affected others and especially Father is a main agent in this... The feelings revealed are so wholesome as well as deep, & there is a wholesome reticence in expression. I think if Mother knew that it could only ennoble him & show the perfectness of married life she would not shrink" (Healey 2001).

FIGURE 6.6 Seventy-nine-year-old Henrietta Darwin Litchfield and her three-year-old grandnephew Richard Darwin Keynes, in 1922, in Henrietta's garden. Professor Richard Darwin Keynes, F.R.S. 1959, carried the continuous Darwin gene pool of those becoming Fellows of the Royal Society—which began with Dr. Erasmus Darwin, F.R.S. 1761—into the sixth generation. Sir Charles Galton Darwin, F.R.S. 1922 (son of George Darwin), was the last individual bearing the Darwin surname to be a Fellow of the Royal Society. He died in 1962, ending the 201-year string of Darwins as Fellows since Erasmus's election.

Henrietta became the family center, much as Emma was, always watching over the activity of her brothers, their children, and their grandchildren. She often read to her nieces and nephews, as Emma had done for Henrietta and her brothers and sisters (Figure 6.6).

In 1915, about 33 years after Darwin's death, a somewhat shady evangelist and temperance preacher, Elizabeth Reid Cotton, also known as Lady Hope, concocted a story that she had visited Darwin at Down House about six months before his death, finding him in bed reading the Bible and regretting his evolutionary ideas. She published this fabrication in the *[Boston] Watchman-Examiner* on 21 October. Her apocryphal story was retold and embellished by Christian fundamentalists until it evolved into the "Darwin's deathbed-conversion myth." This greatly upset the Darwin family. The Darwin children adamantly and persuasively denied that Lady Hope ever visited Darwin. Francis Darwin accused her publicly of dishonesty; she stopped telling her story and never responded to his charges. Henrietta made a very clear statement that should have ended all rumors, but even today fundamentalist preachers continue to use Lady Hope's lie to attract an audience. This topic is treated in depth by Sloan (1960, 1965), Atkins (1974), and Moore (1994).

Here is what Henrietta had to say about the matter, published on page 12 in the 23 February 1922 issue of a religious newspaper, the *Christian*:

> I was present at his deathbed. Lady Hope was not present during his last illness, or any illness. I believe he never even saw her, but in any case she had no influence over him in any department of thought or belief. He never recanted any of his scientific views, either then or earlier. We think the story of his conversion was fabricated in U.S.A. In most of the versions hymn-singing comes in, and a summer-house where the servants and villagers sang hymns to him. There is no such summer-house, and no servants or villagers ever sang hymns to him. The whole story has no foundation whatever. His last words may be found in my book: "Emma Darwin: A Century of Family Letters."

In spite of her many childhood illnesses, Henrietta lived to the ripe old age of 84. There is some discrepancy, even within reliable sources, about the year of Henrietta's death. Freeman (1978) listed 1930, R. Keynes (2005) indicated 1929, and Ogilvie and Harvey (2000) had 1927. Proof comes from Henrietta's obituary, which appeared in the *Times* [London] on 24 December 1927; correct date of her death was 17 December 1927.

7

GEORGE HOWARD DARWIN

(1845–1912)

In April 1845 Charles was hard at work writing his *Geological Observations on South America*, which was published the following year. He was also preparing the second edition of *Journal of Researches* (*The Voyage of the Beagle*), which was published in August 1845 (Freeman 1977).

Charles and Emma's second son was born at Down House on 9 July 1845. He was named George, with Darwin reflecting on pleasant memories of Professor Henslow's son of the same name. George soon became "Georgy" to family members. One of Georgy's earliest childhood memories was watching his father turn nearly blue while taking a cold shower as part of the water cure in the garden shed on a snow-covered winter morning (Browne 1995).

THE GREAT EXPOSITION

In summer 1851, six-year-old Georgy and eight-year-old Etty went with Charles and Emma to the Great Exposition in London and visited the Crystal Palace (Desmond and Moore 1991). This three-story glass house covered over 75,000 square feet (Loy and Loy 2010). The children were restless and Uncle Ras happily entertained them, as he often did throughout their

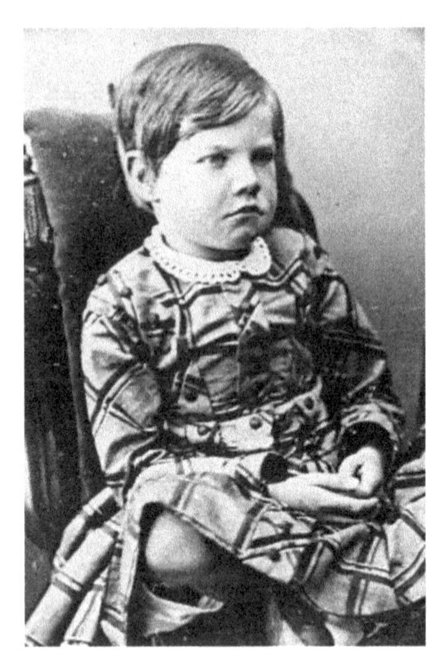

FIGURE 7.1 Six-year-old George Howard Darwin, in August 1851, in the London studio of Richard Beard. The image of Henrietta (fig. 6.1, *left*) was also made during the same photo shoot. Darwin Museum, Down House.

childhood, both at home in Down House and in London. Childfree, Charles and Emma explored the exhibitions from India, Africa, Egypt, Greece, and many other nations. They may have seen an amazing display of John Gould's hummingbirds in the Zoological Society pavilions in Regent's Park (Voss 2010). A drawing of these hummingbirds, and an engraving of Darwin, share the back of the current £10 note, introduced in 2000 by the Bank of England. On another day the Darwin family visited the Zoological Gardens and saw a hippopotamus that was a famous attraction at the time. While in London, George and Henrietta sat for their photographs (Figure 7.1). The trip to London also allowed Charles to safely deliver the first parts of his barnacle monograph to its publisher, the president of the

Ray Society, Edwin Lankester. Charles then called on Weiss and Company to discuss his needs related to dissecting scissors used in his barnacle work (Stott 2003).

DOWN HOUSE CHILDHOOD

Charles was remarkably tolerant of his growing family's interruptions into his research contemplations as the children burst into his study seeking string, pins, paper, ruler, scissors, and other must-have items. George and his older brother William even played cricket in the house (Browne 1995). George recalled how quietly and slowly his father would walk in the woods, a skill honed in the Brazilian rainforest (visited during Charles's H.M.S. *Beagle* voyage), which allowed for the observation of many forms of wildlife.

Charles spent eight years of his life working on barnacles, painstakingly dissecting thousands of specimens in his study from 1846 to 1854 (Stott 2003). To the Darwin children this seemed like a perfectly natural thing for a father to do. One day, while playing with Sir John Lubbock's offspring at their neighbor's property, George asked, "Where does Sir John do his barnacles?" (R. Keynes 2001).

Georgy's informal education began at home. Charles never forced science on his children, but he did everything possible to support whatever interest they showed in any form of natural history. When Georgy wanted to know more about lenses, Charles read a popular optics book to him daily (Browne 1995).

In 1853, England was making preparations for the Crimean War (1854–1856), in which Great Britain, France, the Ottoman Empire, and Sardinia battled Russia. This was the period when Tennyson wrote his poem, "The Charge of the Light Brigade," and Florence Nightingale performed heroic service as a nurse.

The martial excitement did not escape the Darwin children. Emma played rousing tunes on the piano and "Sergeant" Georgy marched around the grounds pretending to be a soldier and commanding his younger brother, "Private" Franky, in mock battles with toy rifles and bayonets (Desmond and Moore 1991). Charles's old shipmate and friend, Lieutenant Bartholomew Sulivan, was now Captain Sulivan. During his visits to Down House, the captain had many heroic tales to tell about various battles. The war, to everyone's relief, ended in 1856 with the Treaty and Declaration of Paris. The allies' victory ensured the integrity of Turkey, Black Sea neutrality, and free navigation of the Danube River.

The Reverend George Varenne Reed was employed as the local tutor for George from the time when the boy was about age seven. The relationship was eventually expanded to include the other Darwin boys, Francis, Leonard, and Horace. Apparently George exhibited maturity and good judgment at a young age, because he was allowed to ride his pony alone on a 20-mile trip to visit relatives in Hartfield, Sussex, when he was only 10 years old (F. Darwin 1916).

CLAPHAM BOARDING SCHOOL

In 1856 George attended nearby Clapham Grammar School to learn math and chemistry from Darwin's contemporary at Cambridge, the Reverend Charles Pritchard. Pritchard, an astronomer and Fellow of the Royal Society, was the founder and headmaster of Clapham. He operated a small astronomical observatory at the school. Darwin recalled his own dislike of the formal, classical education in Latin and Greek that he received, so all of his sons (except William) went to this school. Only George and Francis were taught personally by Pritchard, who eventually became professor of astronomy at Oxford University (Freeman

1978). The training at Clapham was far superior to most education in what Americans call public schools. Pritchard's astronomical influence is clearly visible in George's career.

At Clapham George caught measles, a dangerous disease at that time (Browne 2002). This became one more health issue for Charles and Emma to worry about. In 1861, 16-year-old George had to return home from Clapham in order to have dental surgery (Figure 7.2). Charles took his son to a London dentist, Mr. Woodhouse. Fortunately, chloroform was now in use, and George had to be put under twice in order to treat the extensive decay and remove the offending teeth (Loy and Loy 2010). During home visits, George would join in his father's latest obsession, billiards. He also played cricket in local games against villagers and collected moths for his father.

CAMBRIDGE UNIVERSITY

George entered Trinity College of Cambridge University in 1864 without a scholarship, but by 1866 he had won one. Although he received a degree in mathematics, his skills were slow to develop. According to one of George's classmates, "he took his studies lightly so that they did not interfere with his enjoyment of other things" (F. Darwin 1916). Nonetheless, he exceeded his father's expectations and, at age 23, finished as the second-highest scorer ("second wrangler") on the Mathematical Tripos (a written examination covering the mathematics course taught at Cambridge University). Later that year he was elected a fellow of Trinity College. He seemed to have an ability to focus intently on things that interested him. Charles had no idea that George was so mathematically capable, and wrote him a letter from Down House on 24 January 1868 (Burkhardt et al. 1985–, 16, pt. 1: 33):

FIGURE 7.2 Seventeen-year-old George Howard Darwin, about 1862, photographed by S. J. Wiseman, Southampton. Cambridge University Library. *Right*: George Howard Darwin, photographed by A. Nicholls, Cambridge. Cambridge University Library.

My Dear Old Fellow,

I am so pleased. I congratulate you with all my heart and soul. I always said from your early days that such energy, perseverance and talent as yours would be sure to succeed; but I never expected such brilliant success as this. Again and again I congratulate you. But you have made my hand tremble so I can hardly write. The telegraph came here at eleven. We have written to W. and the boys.

God bless you, my dear old fellow—may your life so continue.

Your affectionate Father,
Ch. Darwin.

George officially received his B.A. on 25 January 1868. Arthur James Balfour, who became prime minister from 1902 to 1905, was among George's Cambridge friends and tennis mates. George went to Paris in spring 1869 to work on his French skills. Upon his return he was offered a science position at Eton (a prestigious British boarding school) but instead decided to study for the bar, which he did from 1869 to 1872 in London. George was admitted to the bar in 1874 but never practiced law. Ill health, in the form of the usual Darwinian digestive problems, intervened. Like his father, he tried the treatments at Malvern, but to no avail. He was also attended to by Dr. Andrew Clark, who treated Charles and Thomas Henry Huxley (Loy and Loy 2010). George chose a life in mathematics and returned to Trinity College in 1873, remaining there for his entire career.

FATHER'S HELPER

George functioned as his father's ambassador when he took his sister Elizabeth on a visit to Paris in 1869. Charles asked him to call upon various scientific colleagues and present his respects. Charles also sent George and Francis to the United States to liaise with colonial Darwinians. George was tutored in engraving by George Brettingham Sowerby II, who drew all of the illustrations for Darwin's book on barnacles (Browne 2002). Both George and Francis provided illustrations for *Insectivorous Plants*, published on 2 July 1875. George helped beautifully illustrate *Climbing Plants*, which appeared in book form in September 1875, and *The Power of Movement in Plants* (1880) (Figure 7.3).

William, George, and Francis collected earthworms for their father, both locally and when they went on holiday in various places. George also reviewed Darwin's calculations of how much soil earthworms could bring to the surface in a year, which turned out to be, on average, a staggering 10 tons of dry earth per acre (C. Darwin 1881). They concluded that a Roman villa found in a field in Surrey was being submerged by earthworm activity at the rate of one inch every 12 years.

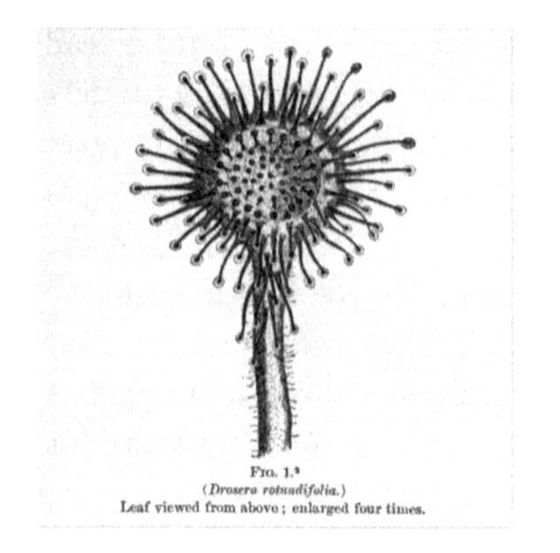

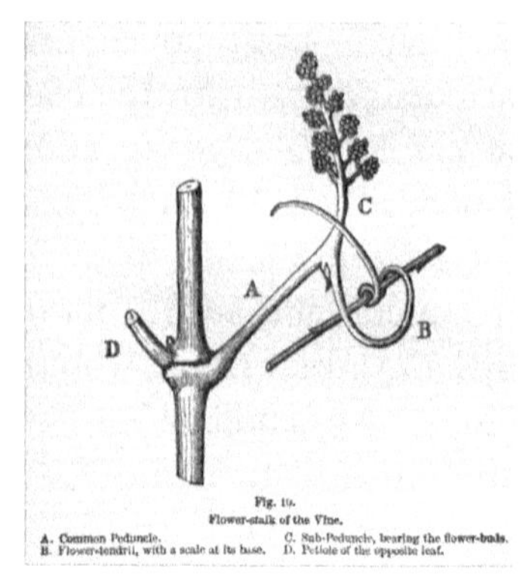

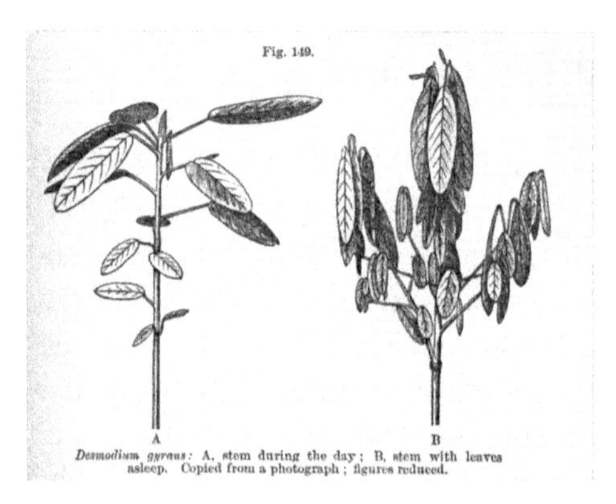

FIGURE 7.3 *Top left*: *Drosera rotundifolia*, drawn by George Darwin for his father's book *Insectivorous Plants*, published in July 1875. *Top right*: Flower-stalk of the vine, drawn by George Darwin for his father's book *Climbing Plants*, published in September 1875. *Bottom*: *Desmodium gyrans* during the day and "asleep" at night, drawn by George Darwin for his father's book *The Power of Movement in Plants*, published in November 1880.

SEANCE

Hensleigh Wedgwood, Emma's brother (Charles's first cousin and brother-in-law) was involved in spiritualism and enjoyed participating in seances. After much struggle he finally got Darwin and Huxley, both notorious skeptics, to attend a seance. George arranged for this performance by medium Charles Williams to take place at Erasmus's house in London on 16 January 1874. It was to be a big social occasion. In attendance were George Eliot (pen name of Mary Ann Cross, aka Mary Ann Evans) and her common-law husband George Henry Lewes, Thomas H. Huxley, the Litchfields, Charles and Emma, Francis Galton, Hensleigh and Fanny Wedgwood, Erasmus Alvey Darwin, and several others. Charles, Lewes, and Evans left before the fun started. The others had a good time and even arranged a second performance at Hensleigh's house a few weeks later. Huxley considered Williams to be a con man, and even learned some of the tricks of the trade to better understand what was done. George John Romanes eventually exposed Williams as a charlatan (Browne 2002). Darwin pronounced it all "rubbish."

NATURE VERSUS NURTURE

Francis Galton, Darwin's half first cousin (Berra et al. 2010a, 2010b), was the author of *Hereditary Genius* (1869) and the founder of eugenics (Pearson 1914–1930). He met with Charles on 2 April 1872 and sent Darwin a survey request on education and background, to which Charles responded on 28 May 1873. His answers to Galton's questions can be seen in *The Life and Letters of Charles Darwin* (F. Darwin 1887, 3: 177–179). In his *Autobiography* Charles wrote, "I am inclined to agree with Francis Galton in believing that education and environment produce only a small effect on the mind of anyone, and that most of our qualities are innate."

Like most of the Darwin household, George was frequently ill as a child and a young man. He spent a lot of time either in bed or undergoing treatments at various spas. This scuttled his legal studies. When back at Cambridge, George began writing essays in order to establish himself. In an article in the *Contemporary Review* (1873), George endorsed Galton's plan for a family register. He also advocated legal divorce in cases of insanity, criminality, and other defects that were considered to be hereditary. Charles supported George's essay. His plant-breeding experiments (C. Darwin, 1862, 1876, 1877a) reinforced Charles's concern about the possible hereditary effects of his own consanguineous marriage to Emma, his first cousin, and its potential link to the ill health of their children (R. Keynes 2001; Berra et al. 2010a, 2010b). He expressed this fear in several letters to his cousin William Darwin Fox and to Joseph Dalton Hooker in 1852–1853 (Burkhardt et al. 1985–, 5: 84, 100, 147, 194). An example of what Darwin was worried about can be seen in his work with yellow toadflax (*Linaria vulgaris*). The five best seedpods were collected from plants that Darwin grew under gauze netting, to prevent cross-pollination by insects. These pods yielded a total of 118 seeds. A single seedpod of a cross-pollinated (uncovered) plant contained 166 seeds. When germinated, the seedlings from the cross-pollinated plants were taller and more robust than the self-pollinated group (Ayres 2008). This made Darwin think about the general ill health of his own offspring who, if they were plants, would fall into the self-pollinated group (Darwin 1876).

MIVART'S POLEMICS

Anti-Darwinist St. George Jackson Mivart saw George Darwin's essay calling for a liberalization of divorce law as an attempt to weaken marriage. Mivart bitterly and unfairly attacked George as a way to get to Charles. Mivart, who was already an enemy of Darwin—demonstrated by his scathing reviews of *The Descent of Man* (Mivart 1871a, 1871b)—used this excuse to continue his attacks on Charles Darwin and *The Descent of Man* in an anonymous article in the *Quarterly Review* (Mivart 1874a; Loy and Loy 2010). Mivart tried to destroy George's reputation as a gentleman by portraying him as licentious. This outraged Charles, who forced Mivart's publisher, John Murray (who also happened to be Darwin's publisher) to publish George's reply (G. Darwin 1874). Printed below George's letter was Mivart's (1874b) anonymous "non-apology." The Darwins contemplated a lawsuit. Thomas Henry Huxley, ever "Darwin's Bulldog," rose to George's defense (Desmond 1997). Some of Mivart's writings were placed on the Vatican's *Index* of forbidden readings, and he was eventually excommunicated from the Roman Catholic church in 1900 by Cardinal Vaughan, mainly due to Mivart's views on hell. He also became persona non grata among the scientific establishment. No love was lost between Darwin's circle of supporters and Mivart. Many years later, Huxley and Hooker continued to block Mivart's application to join the Athenaeum Club, and Charles never spoke to him again (Desmond and Moore 1991).

Charles urged caution, however, in George's concomitant desire to ridicule prayer, divine influence on morality, and heaven and hell. These religious beliefs were not important to Charles, but he had always endeavored to avoid unnecessarily stirring the pot of public outrage, as that would reflect unfavorably on the family.

COUSIN MARRIAGES

Charles began revisions for the second edition of *The Descent of Man* in November 1873. Emma persuaded him to ask George for assistance in this sizeable task. Charles was thinking of paying Alfred Russel Wallace to help with the changes, but he agreed to let George do it (Browne 2002). The second edition of the book was published in autumn 1874.

George did a statistical analysis of cousin marriages via questionnaires (G. Darwin 1875) and concluded that they were very common and not especially harmful (Berra et al. 2010b). Charles took some comfort in this result. George's analysis was one of the first data-based sociological studies (Kuper 2009). George demonstrated "that the percentage of offspring of first-cousin marriages [in mental asylums] is so nearly that of such marriages in the general population, that one can only draw the negative conclusion that, as far as insanity and idiocy go, no evil has been shown to accrue from consanguineous marriages." He also introduced the idea of using the frequency of occurrence of the same surname in married couples (isonymy) to study the level of inbreeding in a population. This concept is still used today in human population biology.

GENEALOGY

As a child George was interested in heraldry (coats of arms, castles, knights, armor, horses, etc.). He pursued genealogy into adulthood, becoming an expert on the family's history and preparing pedigrees for Francis Galton. He hired an American genealogist living in London, Colonel Joseph Lemuel Chester, to produce a pedigree that traced the Darwins back 200 years (Berra et al. 2010b). This document was never printed, but some of Chester's notes are in the Francis Galton archive in the library of University College, London (Freeman 1984).

GEORGE AND LORD KELVIN

Charles Darwin knew that for evolution to produce the great diversity of life that surrounds us today, the earth needed to be immensely old. A professor at the University of Glasgow, William Thompson (Lord Kelvin), whose name is synonymous with thermodynamics, was one of the most distinguished astronomical physicists of the day. His calculations for the age of the earth (estimated at various times to be from 20 to 400 million years, but centering on 100 million years [Dalrymple 1991, table 2.10]) did not provide enough time for evolution to work. Thompson based this estimate on the cooling of the earth (Burchfield 1975; Dalrymple 1991; Kushner 1993). Charles enlisted George, asking him to use his freshly earned math degree to critique Thompson's estimates. As the science of geology and knowledge of radioactivity progressed, Thompson's assumptions were eventually shown to be incorrect.

George's first major scientific paper was published in the *Philosophical Transactions of the Royal Society* in 1877. It dealt with the effects of geological changes on the earth's axis of rotation and how this affected glacial periods. He treated the obliquity of the ecliptic as constant, but suggested that the poles may wander a bit. George was the first scientist to treat the earth as a viscous body instead of either an elastic one (as Thompson had done) or a rigid body (as mathematicians and astronomers had proposed). This made the study of tides much more accessible (Kushner 2004). Lord Kelvin (William Thompson) was the reviewer. Greatly impressed by this paper, he invited George to Glasgow. This resulted in a firm friendship, despite the disagreement between Lord Kelvin and George's father over the age of the earth. Still, George was in a bit of an awkward position as the son of Britain's greatest biologist and the protégé of Britain's greatest physicist (Kushner 1993). By 1897, however, based on George's work with the tidal retardation of the earth's rotation, Lord Kelvin was able to accept earth's age as 1,000 million (1 billion) years old (Dalrymple 1991, table 2.10). Today we recognize that the earth is about 4,600 million (4.6 billion) years old.

Thompson and his wife eventually became godparents to George's first son, Charles Galton Darwin. George considered that his entire scientific career was the outcome of his conversations with Lord Kelvin in 1877, when George was 32 years old (E. Brown 1916). Dalrymple (1991) provided a technical and historical review of Lord Kelvin's and George Darwin's contributions to age-of-earth studies.

HONORARY DEGREE FOR CHARLES DARWIN

Since George was at Cambridge, he was the person who first notified his father that Charles was to be awarded an honorary degree. When Charles was awarded an L.L.D. (Doctor of Laws) by his alma mater, Cambridge University, on 17 November 1877, his sons George, Francis, and Horace, as well as Emma and daughter Bessy, were in attendance. The *Cambridge Chronicle* of 24 November 1877 reported that when Darwin appeared in his red academic gown, he received a huge ovation; a stuffed monkey, also dressed in academic regalia, was released on cords stretched across the gallery.

Two years earlier, Charles had been elected a foreign associate of the Royal Academy of Sciences at Rome. This tickled anti-papist Emma, who was pleased that this was accomplished under Pope Pius IX's nose (Loy and Loy 2010).

TIDES

In papers in the *Philosophical Transactions of the Royal Society* from 1879 to 1881, and especially in 1880, George proposed his theory of the tidal evolution of the earth-moon system, which eventually led to his fission hypothesis for the origin of the moon. A rotating, self-gravitating mass like the earth would assume a pear shape. As its rotation rate increased, an unstable bifurcation could occur. George reasoned that this could split apart (fission) into the earth-moon system. He suggested that the gravitational attraction of the sun produced tidal oscillations in the crust and upper mantle early in the earth's formation, which pulled out a chunk of the rapidly spinning and elongating body. George's work on this topic also affected many areas of mathematics. For example, his research into the equilibrium of rotating fluid masses clarified the pear-shaped figure of equilibrium advocated by Henri Poincaré (Kushner 1993).

Osmond Fisher, a geologist, then suggested that the Pacific Ocean basin was created when the moon was ripped from the earth (O. Fisher 1881). Although this Darwin-Fisher (or extraction) model persisted for the next half century, information gleaned from seismology, plate-tectonics theory, the analyses of lunar rocks brought back by astronauts, and other data do not support this idea. The current view of lunar formation, referred to as the giant impact theory, was proposed by Hartman and Davis (1975), who suggested that 4.5 billion years ago, during the early formation of the earth, a small planetary body collided with it. Some of the resulting debris coalesced in orbit around the earth and became our moon.

George's papers explained that the tides on the earth's surface were caused by the moon and the sun, and that these tides, in turn, affected the moon (Kushner 2004). He demonstrated that lunar tidal friction was the cause of the lengthening of the day (Larmor 1912). On 29 October 1878, Charles wrote to George, "Hurrah for the bowels of the earth and their viscosity and for the moon and for the Heavenly bodies and for my son George (F.R.S. very soon)" (F. Darwin 1916).

George's early papers (1877–1880s) launched his career into the upper tier of scientists of the day. He considered Lord Kelvin to be his master and colleague. He was, as his father had proudly predicted, elected a Fellow of the Royal Society in June 1879, an honor bestowed on his father 40 years previously.

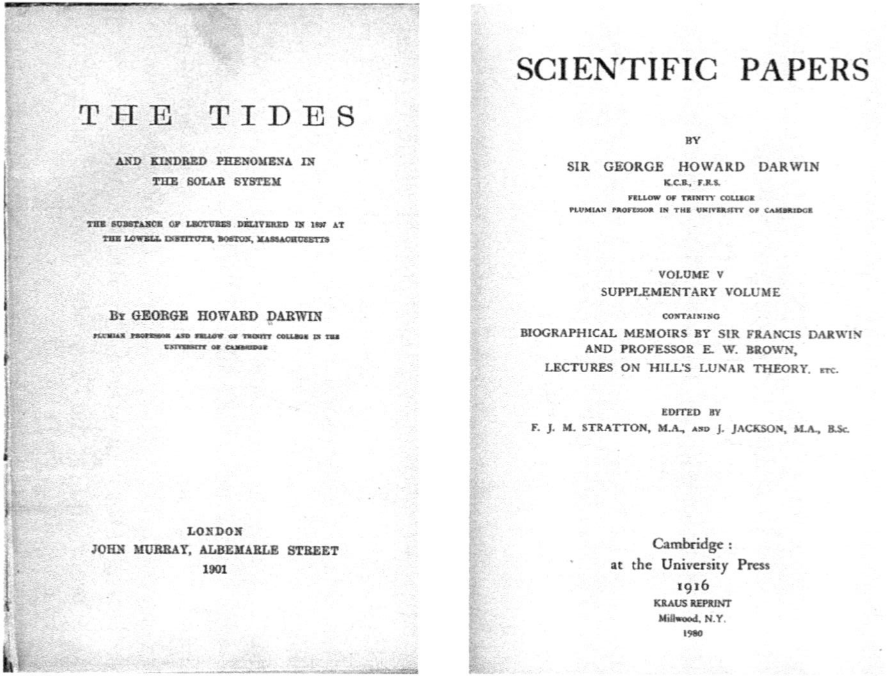

THE TIDES

AND KINDRED PHENOMENA IN
THE SOLAR SYSTEM

THE SUBSTANCE OF LECTURES DELIVERED IN 1897 AT
THE LOWELL INSTITUTE, BOSTON, MASSACHUSETTS

BY GEORGE HOWARD DARWIN
PLUMIAN PROFESSOR AND FELLOW OF TRINITY COLLEGE IN THE
UNIVERSITY OF CAMBRIDGE

LONDON
JOHN MURRAY, ALBEMARLE STREET
1901

SCIENTIFIC PAPERS

BY

SIR GEORGE HOWARD DARWIN
K.C.B., F.R.S.
FELLOW OF TRINITY COLLEGE
PLUMIAN PROFESSOR IN THE UNIVERSITY OF CAMBRIDGE

VOLUME V
SUPPLEMENTARY VOLUME
CONTAINING
BIOGRAPHICAL MEMOIRS BY SIR FRANCIS DARWIN
AND PROFESSOR E. W. BROWN,
LECTURES ON HILL'S LUNAR THEORY, ETC.

EDITED BY
F. J. M. STRATTON, M.A., AND J. JACKSON, M.A., B.Sc.

Cambridge:
at the University Press
1916
KRAUS REPRINT
Millwood, N.Y.
1980

FIGURE 7.4 *Left*: The title page of the second edition of *The Tides*. The first edition was published in 1889. *Right*: The title page of volume 5 of George Darwin's *Scientific Papers*, which features a biographical sketch by his brother Francis Darwin and George's former student E. W. Brown.

George became the government's expert on tidal observations throughout the empire, and the world's foremost authority on tides. He authored the article on "tide" for various editions of the *Encyclopaedia Britannica*, including the eleventh edition (1911), called the scholar's edition. Like all of George's scientific work, the encyclopedia article is very complicated, densely mathematical, and difficult for the nonspecialist to follow. *The Tides and Kindred Phenomena in the Solar System* (G. Darwin 1898), destined for a more general audience, was based on lectures George gave in Boston at the Lowell Institute in 1878. This book was a bestseller and widely translated (Figure 7.4). George served on the Meteorological Council and exerted a powerful influence on all areas of official meteorology (F. Darwin 1916). His tidal predictions were made available worldwide, via reports to the British Association from 1883 onward (Larmor 1912). George's focus was to apply the techniques and methods of mathematical physics to the concerns of geology.

CHARLES DARWIN'S FUNERAL

George's flurry of scholarship came to an abrupt, temporary halt as his father approached death. George traveled to Down House on 10 April 1882 to help Frank support their dying father. George returned to Cambridge on 18 April, and Charles passed away the next day. George hurried home again to help with the funeral and organize pallbearers. William and George made most of the funeral arrangements (Desmond and Moore 1991).

MARRIAGE TO MAUD

In autumn 1882, 37-year-old George proposed to Nellie du Puy, who was 18 years old at the time. She refused George's proposal, but he was soon to get a second shot at a du Puy female. Twenty-two-year-old Martha Haskins ("Maude") du Puy of Philadelphia, Pennsylvania, visited Cambridge in the spring of 1883 to see the same family and friends who introduced her sister to George. The du Puy family was descended from French Huguenot stock that migrated to New York in 1713. Gwen Raverat (1952) (Maud's daughter) delightfully described the social scene and courtship. Gwen also provided her mother's description of her father upon seeing him for the first time, the day after her arrival in Cambridge: "Jane came to tell me that Aunt C. [Cara] wished me to come downstairs to met Mr. Darwin. I ran down and opened the door quickly before I could lose courage, and G. D. quickly stepped forward blushing rosy-red and shook hands. The first thing that struck me was his size. He is little [he was actually 5 feet 10 inches tall]. He is intensely nervous, cannot sit still a minute. He is full of fun, and talks differently from an American man. They are so different in everything." Maud liked English gentlemen, but thought that they were strange and cold. She preferred them to American men, however, because Englishmen "read more and think more, and know more."

Several suitors pursued Maud during her time in Cambridge and one proposed. George was consulted on the suitability of this match. He approved, but Maud had other ideas. George and Maud spent more time together during family activities, and a friendship ensued as Maud began to develop feelings for George. Maud, in the company of another aunt, traveled to continental Europe to see the sights. George followed her, and they became engaged in Florence in March 1884. They were married on 22 July 1884 in Erie, Pennsylvania. George was the last of the bachelor Darwin boys (Loy and Loy 2010).

The newlyweds settled into a house that George purchased in 1885, Newnham Grange (now called Darwin College) in Cambridge. They remained in that yellow brick house near the river for the rest of George's life. The estate had granaries on it, and a mill was nearby. Details of the day-to-day life of the George Darwin household are lovingly told by his daughter, Gwen Raverat (1952). George and Maud had five children, one of whom died in childhood. Gwen, who married Jacques Raverat, was an artist and writer; Charles Galton Darwin became a mathematician and professor like his father; Margaret Elizabeth married Sir Geoffrey Keynes and wrote a biographical sketch of Leonard Darwin; and William Robert was in the army.

DEATH OF EMMA DARWIN

George and Maud and their children often visited Emma at Down House, along with Henrietta. They would amble along the Sandwalk and explore childhood places that held fond memories. Most of Emma's married life was spent at Down House. In autumn 1883, a year and a half after Charles's death, Emma took a house (named the Grove) on Huntington Road in Cambridge to be near George, Francis, and Horace (Healey 2001). The plan was for Emma and Bessy to spend winters in the Grove and summers in Down House. Emma continued to enjoy her children and grandchildren. She was still able to play the piano for them in the last year of her life, even though she was losing her hearing (Loy and Loy 2010). Back in Cambridge, George visited his mother nearly every day. Emma died quietly in Cambridge on 2 October 1896, at the age of 88. As a result of 43 years of marriage to Charles and exposure to his family and scientific friends, Emma had become more secular and humanistic and less rigidly religious (Loy and Loy 2010). She was buried in the churchyard at Downe. George had been Emma's favorite son since childhood and was very attached to Down House. Emma's will specified that George would inherit Down when she died (Healey 2001).

CAMBRIDGE PROFESSOR

George was appointed Plumian Professor of Astronomy and Experimental Philosophy at Cambridge in 1883, and his teaching duties involved lecturing on advanced mathematics. His students loved him and considered him a lifelong friend. They felt free to visit him in his home. Former pupil E. Brown (1916) wrote, "To have spent an hour or two with him, whether in discussion on 'shop' or in general conversation, was always a lasting inspiration." Eventually his students came to dominate the leading astronomical positions and professorships (Kushner 1993).

George was instrumental in the 1899 formation of the Cambridge University Association, which promoted the university's financial affairs among its graduates throughout the world (Larmor 1912). This is what Americans would recognize as an alumni association that works to boost the university's endowment fund.

In 1883 George was awarded the Telford Medal of the Institution of Civil Engineers, and in 1884 the Royal Society bestowed the Royal Medal on him, which he called a wedding gift, since it was presented in the same year as his marriage (Kushner 2004). As Kushner (1993) put it, "by the early 1880s [George] Darwin had himself become a member of those upper echelons of the scientific establishment—aided at first by familial ties, abetted by the powerful patronage of Thompson, but ultimately propelled there by the significance of his contributions."

SIR GEORGE—WORLD-CLASS AUTHORITY IN GEOPHYSICS

Several series of highly technical, major papers arose out of George's theory of tidal evolution, which led to him being considered Britian's leading geodesist. His work resulted in an explosive growth of the geophysical sciences, and George Darwin's reputation is known by multiple generations of geophysicists (Kushner 2004). He was president of the Royal Astronomical Society (1899–1900), twice president of the Cambridge Philosophical Society (1890–1892 and 1911–1912), and president of the British Association for the Advancement of Science. He became a knight commander of the Order of the Bath in 1905. This honor was especially important, as it was announced to him by his college friend, Prime Minister Balfour. The knighthood and Cambridge University Press's publication of his collected *Scientific Papers* in five volumes (1907–1916) were among the most treasured events in George's life.

The five volumes of *Scientific Papers* included over 80 separate titles (Figure 7.4). The titles of the five volumes give a brief overview of the depth and breadth of his research: volume 1 (1907), *Oceanic Tides and Lunar Disturbance of Gravity*; volume 2 (1908), *Tidal Friction and Cosmogony*; volume 3 (1910), *Figures of Equilibrium of Rotating Liquid and Geophysical Investigations*; volume 4 (1911), *Periodic Orbits and Miscellaneous Papers*; and volume 5 (1916), *Supplementary Volume*. They reflect George's career as an applied mathematician who tested theories of cosmogony and had a preference for quantitative rather than qualitative results. E. Brown (1916) provided an analysis of George's scientific work in all its excruciating detail. Kushner's (1993) account is much more understandable. He wrote, "Finally, it must be admitted that [George] Darwin's method is plainly intimidating to the mathematically illiterate, and its pages of formulas forbidding even to the mathematically sophisticated: even highly qualified referees such as Thompson and Lord Rayleigh often read through Darwin's dense mathematical prose in a cursory manner."

In 1911 George received Britain's highest scientific distinction, the Copley Medal of the Royal Society. His father had received this award in 1864. Francis Galton, George's half first cousin once removed, received the Copley medal in 1910, with strong support from George, who was unaware that his own name was on the list for recognition (Larmor 1912). George received doctorates from 9 universities, held foreign memberships in 18 academies, and was a correspondent in 10 foreign academies. He clearly occupied a central position in the scientific aristocracy. Larmor (1912) referred to George as the "doyen of the mathematical school at Cambridge." Kushner (1993) was willing to call him the "head of the British school of geophysics." Sir Harold Jeffreys dedicated his book, *The Earth* (1924), to the memory of Sir George Howard Darwin, "father of modern geophysics and cosmogony."

DAUGHTER GWEN RAVERAT, FAMILY BIOGRAPHER

Two of the very best books about Charles Darwin and the family were written by George's descendants: his daughter Gwen Raverat (*Period Piece*) and great-grandson Randal Keynes (*Annie's Box*). Gwen reminisced about her father and her relatives. George was of average height and slightly built, at about 137 pounds. Besides science, he loved history, language, travel, and words. He was a world traveler, visiting many European countries, America, South Africa, and other exotic places. He had to give up tennis after being hit in the eye by a ball in 1895. He wanted to know the proper terminology for everything he did, such as archery, which he took up in 1910. For example, what was the proper term for the notch in the arrow's posterior? Answer, a nock. George was proficient enough to win several local archery medals and trophies. He had a romantic streak, which spurred his lifelong interest in heraldry, and he

FIGURE 7.5 Sir George Howard Darwin, from a watercolor drawing by his daughter, Gwen Raverat. This reproduction is from the frontispiece of volume 5 of *Scientific Papers*, reprint edition, 1980.

was delighted at having been made a knight of the Bath. He read widely, including Shakespeare, Chaucer, and works on prehistory (Figure 7.5). Gwen described her father as the most worldly, active, and alert of the Darwin brothers, her uncles. George was genuinely fond and considerate of people and completely unselfish. He seemed to be surprised by his own success.

Examples of George's thoughtfulness and generosity include raising money for a railway guard who was robbed of his savings, and intervening with authorities to make sure a one-legged man was sent to London to be fitted with an artificial limb (F. Darwin 1916).

DARWIN FAMILY NAME

George died of pancreatic cancer on 7 December 1912 at Newnham Grange in Cambridge, at the age of 67. He was buried at Trumpington. George was recognized as the world's leading authority on tides and one of the most eminent scientists of late Victorian and Edwardian Britain (Kushner 1993).

Maud, 16 years younger than George, lived until 1947. The only remaining male lineage of Charles Darwin's family (and hence the Darwin name) passes through George Darwin.

ELIZABETH DARWIN (1847–1926)

By 1847 Charles was well into his barnacle research. He attended several meetings of the Geological Society Council in January, April, May, and June, as well as the British Association for the Advancement of Science meeting in Oxford in June. This rather ambitious (by Darwin's standards) scientific/social schedule was briefly interrupted when, on 8 July 1847, Elizabeth Darwin put in her first appearance at Down House.

AN UNUSUAL CHILD

Her nickname in childhood was "Lizzy," but she opted for "Bessy" as she matured. Annie's nickname for her baby sister was "little duck" (R. Keynes 2001). At the end of 1849, two-year-old Lizzy came down with scarlet fever after Annie and Etty were infected. Lizzy and Etty recovered, but Annie remained weak. Lizzy and Georgy had birthdays on successive days in July, and Uncle Ras would show up at Down House for their amusement on such celebratory occasions.

Etty and Lizzy were four years apart and interacted with the female staff of Down House (cook, governess, nurse) while the boys (Georgy, Franky, Lenny, and Horace) roughhoused outside. Access to female cousins helped somewhat, but Lizzy, as the youngest girl, was rather isolated. By the age of five, she seemed to use and pronounce words in a strange way, and she began talking to herself while seeking privacy (Figure 8.1).

FIGURE 8.1 Elizabeth "Bessy" Darwin, in the mid-1850s. English Heritage Photographic Library, London.

Lizzy did not appreciate being interrupted when doing this (Desmond and Moore 1991). She shivered, made grimaces, and was unusually quiet for a child. She had an odd way of moving her fingers when excited, similar to Charles's own nervous tic. This similarity emphasized his view that his children had inherited his medical problems (R. Keynes 2001). Much later, in January 1868, when Charles was nearly 59 years old and Lizzy was 21, he gave a very nuanced and thinly disguised account of his and Lizzy's tic in *The Variation of Animals and Plants under Domestication* (2: 6–7 in the first edition; 1: 450–451 in the second edition):

> I will give one instance which has fallen under my own observation, and which is curious from being a trick associated with a

> peculiar state of mind, namely, pleasurable emotion. A boy had the singular habit, when pleased, of rapidly moving his fingers parallel to each other, and, when much excited, of raising both hands, with the fingers still moving, to the sides of his face on a level with the eyes; when this boy was almost an old man, he could still hardly resist this trick when much pleased, but from its absurdity concealed it. He had eight children. Of these, a girl, when pleased, at the age of four and a half years, moved her fingers in exactly the same way, and what is still odder, when much excited, she raised both her hands, with her fingers still moving, to the sides of her face, in exactly the same manner as her father had done, and sometimes even still continued to do so when alone. I never heard of any one, excepting this one man and his little daughter, who had this strange habit; and certainly imitation was in this instance out of the question.

Charles thought that Lizzy was not imitating an action of his that she had observed, since he had never been aware of her having seen him do it.

Because of Lizzy's eccentricities, Emma and Charles were very protective of their unusual daughter. She was home-schooled in languages, geography, and music, with a smattering of simple botany, but no science or mathematics. Beginning in January 1863, Lizzy did spend a few months at Miss Buob's boarding school in Kensington, at her own request (Loy and Loy 2010). This helped her achieve a modicum of independence and self confidence. Browne (1995) suggested that her clumsiness and speech patterns may have been related to mild cerebral palsy, but also pointed out that Lizzy's letters reflected a very alert and clever mind. Her speech oddities seemed to have disappeared by about age six (Loy and Loy 2010). As a child, Lizzy was sometimes described as not very bright and a little difficult (Healey 2001), but with the loving care and encouragement of Emma she grew into a young woman who helped Emma with household duties and with the care of Charles during his illnesses. Like many in the Darwin household, Lizzy had her share of dental problems. At age 13, she had two teeth filled and three extracted (Loy and Loy 2010).

Francis's son Bernard, who was raised in Down House after the early death of his mother, knew his Aunt Bessy better than most of the family, since she had helped Emma raise Bernard. Both of them doted on the boy. He considered Aunt Bessy to be very sensitive, and an instinctive judge of character, even if she was a bit helpless in managing things and could not grasp percentages (Healey 2001).

THE MOST ENIGMATIC DARWIN

Emma took her daughters to church, but by 1866, 19-year-old Bessy refused to become confirmed (Figure 8.2). She told Emma that she did not believe in the Trinity or baptism and wanted nothing to do with the catechism (Browne 2002).

As the Darwin girls began to assert their independence, Bessy traveled to continental Europe with Henrietta and went to Germany on her own in 1866 (Browne 2002). A family group photograph, including Bessy, circa 1866 reflects the wealth of the Darwin household (Figure 8.3). In 1868 Alfred Russel Wallace and his wife visited Down House. Bessy wrote to Henrietta that he was "very pleasant."

Gwen Raverat loved her Aunt Bessy dearly and provided some affectionate insights into this most unknown and enigmatic of all Darwins, one whose only interest was in her family. Bessy enjoyed reading aloud to Gwen and her brothers and sisters (Bessy's nieces and nephews). Gwen described her as "very stout and nervous, and apt to fumble her fingers when agitated. She was not good at practical things it must be confessed; and

FIGURE 8.2 Elizabeth Darwin, about 18 years old, seated in a window of Down House, photographed by Leonard Darwin, circa 1865. Cambridge University Library.

she could not have managed her own life without a little help and direction now and then; but she was shrewd enough in her own way, and a very good judge of character." Bessy was very loyal to her friends, as well as rather brave to have expressed some skepticism over her sister Henrietta's continual ill health. Gwen described "Aunt Etty as rather superior and impatient; and Aunt Bessy submissive, but a little resentful and critical." Bessy enjoyed life through others, and Gwen wrote that "she was so warm-hearted and dependent on affection, that we found it easy enough to love her." According to Gwen's childhood memories, Bessy didn't do much useful work at Down House. She read a lot (including works in French), went for walks, wrote letters, and sat on the veranda talking with other ladies and knitting with Emma (Figure 8.4). Bessy tried dieting, but without much luck, and by January 1887 she weighed 188 pounds (Loy and Loy 2010).

FIGURE 8.3 Emma Darwin, reading to her children while seated in a bay window at Down House, in this circa 1866 photo. *Left to right*: Leonard, Henrietta, Horace, Emma, Elizabeth, Francis, and a school friend called "Spitta." Cambridge University Library.

FIGURE 8.4 *Left*: Elizabeth Darwin. Cambridge University Library. *Right*: Elizabeth Darwin, in later years. Darwin Museum, Down House.

SPINSTERHOOD

Gradually the Darwin brothers moved from Down House to pursue their educations, careers, and marriages. Henrietta and her husband Richard Litchfield established their own household in London in 1871. Bessy, who never married, remained at Down House with Emma and Charles. She was at her father's bedside when he died in 1882, and she attended his funeral in Westminster Abbey. When Emma moved to Cambridge, Bessy, Frank, and Bernard went with her to the new home, known as the Grove (Wedgwood and Wedgwood 1980). Frank soon bought a nearby house of his own. Emma and Bessy returned to Down House every summer, and the Darwin children and grandchildren gathered there for family visits.

When Emma died in 1896, Bessy became more independent. She bought a house in Cambridge, to be near her brothers, and resided there with a companion (Healey 2001). She ventured abroad, and volunteered at an old folks' home and read to the ladies. She lived long enough to witness Bernard's success as a golfer and writer and to entertain his son and two daughters. She continued to flutter her hands (Healey 2001). Elizabeth died on 8 June 1926, one month shy of her seventy-ninth birthday.

FRANCIS DARWIN 1848–1925

In 1848 Charles Darwin was still deeply immersed in his barnacle dissections. Attendance at several Geological Society Council meetings served as an occasional break from his labors. He also spent time with his dying, 82-year-old father. Francis Darwin was born on 16 August 1848, and 39-year-old Charles was quite unwell during the second half of that year. Francis's birth was within a year of chloroform's discovery as an anesthetic, so Emma's discomfort in childbirth may have been less than usual.

RUSTIC SOUNDS OF DOWN HOUSE

On 8 March 1849 the entire family—six children, including baby Francis—and servants traveled to the Malvern spa of Dr. Gully so Charles could be treated with water therapy for his chronic illness. This was on the recommendation of Charles's friend and former shipmate, Bartholomew Sulivan, now Captain Sulivan, who was worried about Charles's sickly appearance when Bartholomew visited Down House (Desmond and Moore 1991). Darwin remained at Malvern until 30 June. The rest, if not the treatments, did seem to do Charles some good. During this time, Francis was christened at the local priory church.

Young Francis had older brothers Willy and Georgy to run with outside while Etty and Bessy played inside (Figure 9.1). Quite naturally he was dubbed "Franky," and he eventually became "Frank." Eight years before his death, Francis wrote a memoir of his idyllic childhood at Down House, entitled *Rustic*

FIGURE 9.1 Francis Darwin. Cambridge University Library.

Sounds (1917). It included descriptions of seasonal bird songs and human sounds. He described the noises associated with drawing water from a very deep well: the turning of the fly-wheel, and the raising of a bucket on a wire rope. All the Darwin children loved the sound their father made when taking his daily exercise around the Sandwalk. Francis described it as a rhythmical click, click, click as Charles's heavy, iron-shod walking stick struck the ground. A Down House gardener, Henry Lettington, taught Franky to make whistles and bird traps and helped him with his pet rabbits (R. Keynes 2001). Francis recalled the visits of Joseph Dalton Hooker to Down House and how Hooker loved to eat gooseberries in the kitchen with the children. Other childhood memories included how well his father treated the family servants, never raising his voice to them and always speaking politely, and how angry his father got when he observed a villager mistreating a horse or other animal (R. Keynes 2001).

While playing soldier with "Sergeant" Georgy, "Private" Franky shooed away his father at the point of a bayonet when Charles tried to kiss him. Francis later explained that kissing his father did not seem to be in keeping with his military guard duties (F. Darwin 1916).

A SCIENTIFIC CHILDHOOD

In 1856, as Charles pondered the dispersal of plants across salt-water barriers, eight-year-old Franky suggested that a floating dead bird with a crop full of seeds might be a dispersal mechanism (Burkhardt et al. 1985–, 6: 305). The experiment was performed, and viable plants resulted from the seeds in the crop of a pigeon that floated in salt water for 30 days. This is typical of the sort of intellectual atmosphere swirling around the household in which the children grew up. As Francis got older, Charles relied on him for all sort of lab jobs around the house, such as skinning and stuffing animals and tending the experimental plants. When Charles was working too hard, Emma used the boys to scout out and then drag Charles off to some lodge for a holiday. By 1859 Francis (age 11), Leonard (age 9), and Horace (age 8) were all involved in collecting beetles, and Francis recalled how delighted Charles was when the boys discovered a scarce specimen (Burkhardt et al. 1985–, 7: 196).

FROM MEDICINE TO PLANT PHYSIOLOGY

First George, and then Frank and Leonard, were tutored by the Reverend George Varenne Reed before being sent to Clapham School. In 1866 Frank entered Trinity College of Cambridge University without a scholarship (Figure 9.2). He initially studied

FIGURE 9.2 Francis Darwin, photographed by Bassano, London. English Heritage Photo Library.

math, but then switched to natural sciences, in which he obtained a B.A. degree with first-class honors in 1870. In summer 1871 George and Frank travelled to the United States on holiday, both to observe the scientific scene for their father and to call on Asa Gray at Harvard University. After returning to England, Frank moved to London to study medicine at St. George's Medical School, where he earned an M.B. (Bachelor of Medicine) in 1875. Like his Uncle Ras, Francis never practiced medicine. While in London he lived with Uncle Ras, whose social lifestyle was not exactly conducive to Francis's application to his studies. Charles had to nag his son to complete his thesis, "On the Primary Vascular Dilatation in Acute Inflammation" (F. Darwin 1876; *Nature* 1925). With his experience in animal physiology, which

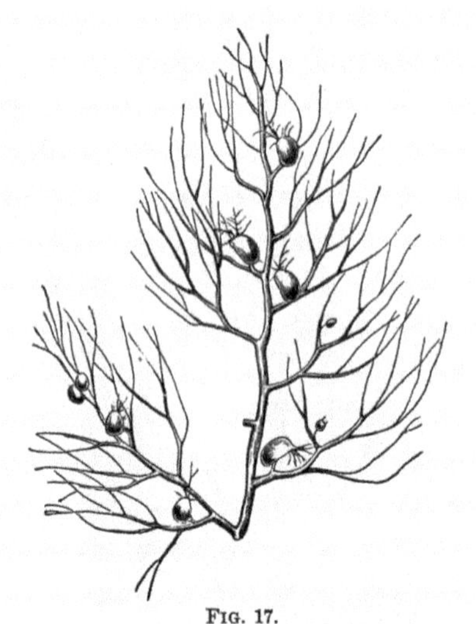

FIGURE 9.3 *Utricularia neglecta*, drawn by Francis Darwin for his father's book *Insectivorous Plants*, published in July 1875.

he learned in medical school, Francis was able to apply similar techniques to plant physiology. He helped move botany from a tool of medicine to a science in its own right. Francis wrote, "It may be that from the study of plant-physiology we can learn something about the machinery of our lives" (Ayres 2008).

A lab was prepared at Down House, and Frank was to help Charles work on his *Insectivorous Plants*, which was published in July 1875 and included lovely illustrations of *Aldrovanda* and *Utricularia* by Francis (Figure 9.3). Darwin's *Climbing Plants* followed in September 1875. In this work he divided climbing plants into three groups: those that climb via stems, via leaves, or via tendrils. The function of climbing is to give the leaf surface access to as much light as possible, thereby outcompeting its neighbors without the expense of developing the elaborate structure of a tree. By now, Francis was well and truly becoming a botanist instead of a medical man.

LOVE AND LOSS

Frank acquired a Welsh girlfriend, Amy Ruck. Amy's brothers were Clapham School friends with Frank and Horace, and Charles and Emma had visited her parents, Lawrence and Mary Ann Ruck, while on vacation in Wales in 1869 (Loy and Loy 2010). Amy, a raven-haired beauty, was originally interested in Francis's younger brother Horace, but he was three years younger than she. Her attention then turned to Frank (Ayres 2008). Amy became one of many correspondents from around the world who provided Charles with information on earthworms; she measured the depth of Welsh wormholes. Frank and Amy were married on 23 July 1874 at All Souls Church in Langholm Place, London, the same church where Robert Browning and Elizabeth Barrett secretly married in 1846. Charles provided a very handsome wedding gift of £5,000 (Ayres 2008). The newlyweds honeymooned on the Continent.

Frank and Amy lived in a house (Down Lodge) in the village of Downe. Charles considered this slender-faced young woman "sweet and gentle." Francis would walk to Down House daily to help his father with botanical experiments and to deal with the huge volume of correspondence Charles received from around the world. This took several hours each day. Francis had become his father's de facto lab assistant and secretary from 1874 to 1882.

A month before the first Darwin grandchild was born, Charles finished the initial draft of his *Autobiography* on 3 August 1876 (Moore 1989). Frank and Amy's baby, Bernard, was born at Down House on 7 September 1876. The labor was difficult and "instruments" (probably meaning forceps) were used for the delivery (Loy and Loy 2010). Amy developed a fever and convulsions and became unconscious, most likely due to puerperal sepsis. Tragically, on 11 September, four days after giving birth, 26-year-old Amy died, with Frank and Charles at her side. Frank was devastated. This loss was worse for him than Annie's death was for his father. Sister Bessy, who had formed a deep bond with Amy, collapsed, and Emma barely kept her composure. Charles, who had been so excited about the birth of a grandson, was also deeply affected. He wrote to Hooker: "I saw her expire at 7 o'clock in the morning.... My dear old friend, I know that you will forgive my pouring out my grief" (R. Keynes 2001). Frank, his brother Horace, and Amy's parents accompanied her body back to northern Wales for burial in her father's village (Loy and Loy 2010).

When Francis returned from the burial, overcome with grief, he moved into Down House, where Bernard was brought up until 1882. Emma, now 69 years old, suddenly had another baby to raise. A wet nurse was employed for Bernard, and Henrietta enthusiastically assumed a maternal attitude toward babysitting her nephew. Emma and Mary Ruck became very good friends and often visited each other's homes. This friendship, based on their grandson and many shared values, lasted into both ladies' old ages (Ayres 2008). Charles wrote a letter to Frank, based on his own experience with grief over Annie's death, advising him that "the only chance of forgetting for short times your dreadful loss" was in "the habit of close mental attention" (R. Keynes 2001).

FATHER'S ASSISTANT

To that end Charles tried to keep Frank occupied by having him work on editing Charles's *Autobiography* and correcting proofs for the second edition of the *Fertilisation of Orchids*, which was published in January 1877. Francis collaborated with his father on *The Different Forms of Flowers on Plants of the Same Species*, which was published in July 1877. Francis relieved his father of many tedious scientific tasks, such as the microscopic examination of specimens. Undoubtedly he increased his ailing father's research output through his constant help. Down House received a two-story remodeling to make a bedroom and study for Frank and to reposition the billiard room (Desmond and Moore 1991). Years later Francis complained about his father's lack of appreciation for heirlooms when he learned Charles had sold a gold watch given to Dr. Robert Waring Darwin, as well as a beautiful rare vase, in order to finance the purchase of the billiard table (Browne 2002).

BERNARD

Emma opened the old nursery for Bernard, and a village girl was hired as a nursemaid. Bernard soon became the darling of the Darwin household and received attention from everyone. When he was three years old the servants taught him to ride a donkey on the Sandwalk and introduced him to cricket. Charles referred to his grandson as "Dubba," and they would take walks together when Bernard was five years old. Charles would time Bernard's tricycle rides with his pocket watch (Browne 2002). Bernard eventually became a prominent sportswriter for the *Times* [London] from 1908 to 1953. As a journalist, he covered golf events around the world. Bernard played in amateur tournaments and captained the British team in America in 1922. In 1934 he became captain of the Royal and Ancient Golf Club of St. Andrews (Ryde 2004). He was also known for his spiffy sartorial sense (Loy and Loy 2010).

GERMAN PLANT PHYSIOLOGY INFLUENCE

Francis traveled to Germany three times between 1878 and 1881 to learn the latest plant physiology techniques in the laboratories of Julius Sachs at the University of Würzburg and Anton de Bary at the University of Strasbourg. Francis's fluency in German was a great help. He used this new knowledge to stimulate and improve his father's botanical researches. Charles, the gentleman-naturalist with great powers of observation, was introduced to the "high-tech" experimental techniques of the day. Frugal Charles balked, however, at purchasing expensive lab equipment and continued to use improvised, homemade gadgets for his experiments, much to Francis's frustration (Browne 2002). Nevertheless, Francis admired his father's simple ingenuity, as well as the quickness and efficiency of his father's movements when working in the lab. He also recognized Charles's deep appreciation for the beauty of a flower (R. Keynes 2001).

Sachs was critical of Charles' and Francis's techniques and said so in print. The relationship ended badly. Of his visit in 1879, Francis wrote: "I published what seemed to me a harmless paper, in which I criticized some of his researches. I wrote to him on the subject but received no answer. Partly on account of his silence and partly to pay a visit to a friend, I travelled to Würzburg. I found Sachs in the Botanical Garden; he seemed to wish to avoid me, but I went up to him and asked him why he was angry with me. He replied: 'The reason is very simple; you know nothing of Botany and you dare to criticize a man like me'.... And that was the last I saw of the great botanist. I was undoubtedly stupid, but I do not think he showed to great advantage in the affair" (Ayres 2008).

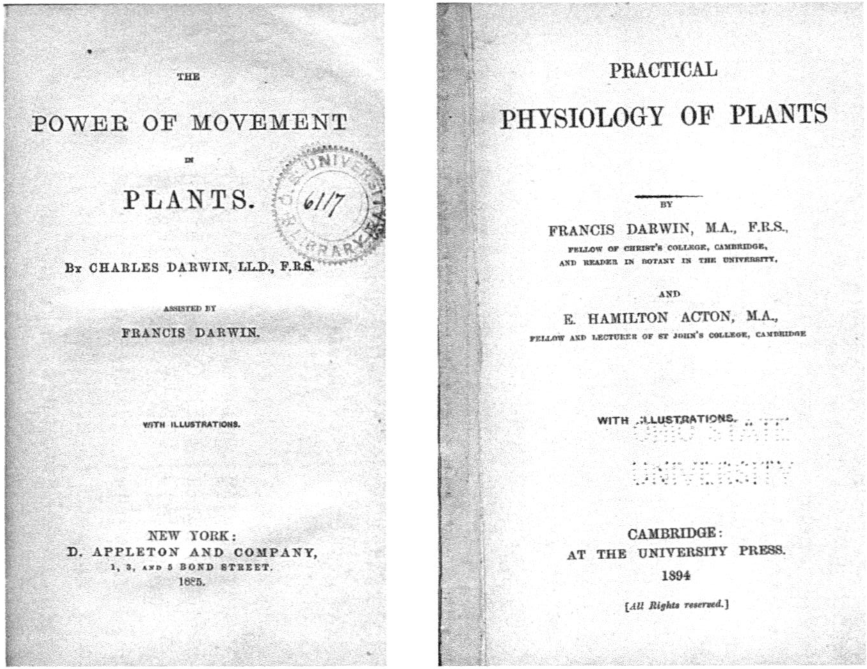

THE

POWER OF MOVEMENT

IN

PLANTS.

BY CHARLES DARWIN, LL.D., F.R.S.

ASSISTED BY

FRANCIS DARWIN.

WITH ILLUSTRATIONS.

NEW YORK:
D. APPLETON AND COMPANY,
1, 3, AND 5 BOND STREET.
1885.

PRACTICAL

PHYSIOLOGY OF PLANTS

BY

FRANCIS DARWIN, M.A., F.R.S.,
FELLOW OF CHRIST'S COLLEGE, CAMBRIDGE,
AND READER IN BOTANY IN THE UNIVERSITY,

AND

E. HAMILTON ACTON, M.A.,
FELLOW AND LECTURER OF ST JOHN'S COLLEGE, CAMBRIDGE

WITH ILLUSTRATIONS.

CAMBRIDGE:
AT THE UNIVERSITY PRESS.
1894

[*All Rights reserved.*]

FIGURE 9.4 *Left*: The title page of an American edition of *The Power of Movement in Plants*. The first British edition was published by John Murray in 1880. *Right*: The title page from the first edition of *Practical Physiology of Plants*, published in 1894. This is one of the first textbooks on plant physiology.

THE POWER OF MOVEMENT

Charles and Francis spent a great deal of time trying to work out how movements occur in the various parts of plants. They concocted all sorts of experiments to catch them at it. The musically inclined Francis even played the bassoon for the plants, to see if they responded to vibrations. They did not. *The Power of Movement in Plants* was eventually published on 22 November 1880, and the title page reads "By Charles Darwin assisted by Francis Darwin" (Figure 9.4). This book is an extension of Darwin's work on climbing plants and a classic in the study of phototropism. It is a pioneering work in the field that would become known as plant physiology, and it is Charles Darwin's most important botanical publication, because of its originality and lasting impact (Ayres 2008). The Darwins researched the effects of gravity, light, and other environmental factors on plant growth and demonstrated that the bending of plant stems was a result of differential growth on one side of the stem compared with the other. This knowledge laid the foundation for the discovery of plant hormones (auxins). The father-son botanists also explored the phenomenon of "sleep" in plants (nyctinasty), where leaves fold together at night (Figure 7.3). Today botanists know that this is a water-conservation mechanism brought about by differential changes in the turgor of cells on opposite sides of part of the leaf stalk (Ayres 2008). Between 1880 and 1908 Frank published a dozen papers on plant growth (Junker 2004).

During the earthworm studies the family joined in to learn if the burrowing creatures could hear. Even Bernard was enlisted to toot his whistle while Frank played the bassoon, Emma played the piano, and Bessy shouted, all to no avail. But when pots of worms were placed on Emma's piano, they did respond to the vibrations. Charles's last book, *Formation of Vegetable Mould through the Action of Worms*, was published on 10 October 1881. The proofs were corrected by Frank.

DARWIN'S DEATH

Father and son were still doing botanical experiments two days before Charles's death. Frank, Henrietta, Elizabeth, and Emma were with Charles when he died on 19 April 1882. Frank, the medical doctor, could not find his father's pulse in those last few moments (R. Keynes 2001). He explained to a questioning Bernard that "Grandpa has been so ill that he won't be ill any more" (Desmond and Moore 1991). It fell to Frank and George to notify Charles's scientific colleagues of his death.

REMARRIAGE AND A DAUGHTER

Francis moved to Cambridge in 1882, after his father died. He lived in one of the houses of what came to be called the Darwinery, because of the multiplicity of well-known residents on Huntingdon Road (*Nature* 1925). Emma and Bessy (the Grove), Frank (Wychfield), and Horace (the Orchard) all had houses there (Raverat 1952). When Emma died, Bessy bought a house of her own in Cambridge.

Francis was made a Fellow of the Royal Society in 1882. The next year, when Bernard was seven years old, Francis married Ellen Wordsworth Crofts, a lecturer in English literature at Newnham College. She gave up the lectureship upon her marriage. Ellen was 27 and Francis was 35 when they married on 13 September 1883. As Francis's son Bernard wrote of his stepmother in his autobiography, "Ellen was always kind as could be in reading to me and playing with me, but there was some feeling of reserve: perhaps she tried too hard to be a good stepmother and never to outstep those limits" (B. Darwin 1955).

Ellen became pregnant in 1884 but suffered a miscarriage, followed by an elevated temperature. The specter of Amy's tragic death was again confronting Frank. The fever broke on the third day, and Ellen eventually recovered (Loy and Loy 2010). On 30 March 1886, Ellen gave birth to a daughter, Frances (Charles and Emma's granddaughter). Frances spent many enjoyable occasions playing with her cousins Gwen (George's daughter) and Nora (Horace's daughter), who were of similar age. All three cousins became famous in their own right. Frances married Professor Francis Macdonald Cornford (who translated Plato) on 1 July 1909 and wrote poetry under her married name, Frances Cornford. Francis Darwin built a house for his daughter and son-in-law on Madingly Road in Cambridge (Ayres 2008).

PLANT PHYSIOLOGY AT CAMBRIDGE

Francis lectured in botany at Cambridge and taught plant physiology with Sydney Howard Vines. When Vines was appointed chair of botany at Oxford, Francis was promoted to reader in botany at Cambridge. He occupied this position from 1888 until 1904. Francis refused to apply for the professorship (chair) of botany when it became vacant, however, saying that a younger man needed it more than he did (Raverat 1952). He preferred to see his old friend Harry Marshall Ward become chair. During this period, Cambridge University included a plant physiology course for undergraduates in its curriculum, before most other universities in Britain even recognized its potential (Ayres 2008). Perhaps Francis felt he was too busy editing his father's letters and checking on his 87-year-old mother to deal with the hassles of chairmanship (Ayres 2008).

WIDOWER AGAIN

When Ellen died in 1903 at the age of 48, Francis, for the second time, was devastated by the loss of a wife he dearly loved. Once again he had a child to raise without a wife. He resigned the readership in botany he had held at Cambridge for 16 years on 28 August 1904, sold Wychfield, and moved to London, but he returned to Cambridge after a year to resume his research.

LEADER OF STOMATAL PHYSIOLOGY

Francis's lengthy paper in *Philosophical Transactions of the Royal Society* in 1898 helped secure his distinction as a pioneer of stomatal physiology (F. Darwin 1898), and his work is still cited to this day by stomatal researchers (Brodribb and McAdam 2011). Stomata are the pores in a leaf surface whose daily cycles of opening and closing regulate the exit of water and the entry of carbon dioxide. Francis also tended to revisions for the second edition of *The Different Forms of Flowers* (1884) and *Insectivorous Plants* (1888). In addition to editing his father's autobiography, letters, and books, Francis continued to work on transpiration and how stomata function to control water loss in plants. He became *the* authority on stomatal physiology (Ayres 2008). He invented several instruments to study this phenomenon, including an ingenious device (a horn hygroscope) made from a shaving of animal horn. It exhibited a strong curvature when one side was exposed to more humid air than the other (*Nature* 1925; Junker 2004). He also invented a porometer that could measure the openness of stomata (Ayres 2008). He coauthored *Practical Physiology of Plants* with E. H. Acton in 1894 (Figure 9.4). Francis's botanical lectures to medical students were published as *The Elements of Botany* in 1895.

GWEN'S DESCRIPTION

Francis's niece, Gwen Raverat (1952), considered her Uncle Frank to be the most cultivated of her uncles, describing him as benevolently philistine. He was a keen musician who played the flute, the oboe, the bassoon, the recorder, and fife and drum. Francis was the only son to share Emma's musical talent. He had a sense of style, a light touch in writing, and an enchanting sense of humor. Gwen considered depression to be his form of the Darwin family hypochondria. Nonetheless, it is reasonable to postulate that Francis's depression owed its origin to the shocking loss of Amy in childbirth. Golf also depressed him, something to which anyone who has ever played the game can relate. Gwen remarked that Frank was the only Darwin son who was a real naturalist, noticing plants and birds just as her grandfather would.

EDITOR OF *LIFE AND LETTERS*

A lot of what we know about the personality and the day-to-day life of Charles Darwin is due to the reminiscences of Francis and his major effort in making his father's letters available in a three-volume work, *The Life and Letters of Charles Darwin* (F. Darwin 1887). This compendium has also gone through several editions that were issued as one- and two-volume publications. It is Francis Darwin's greatest contribution to the history of science. The review in the *Times* [London] from 19 November 1887 was glowing in its praise for Francis's editorial work: "The all-round humanity of [Charles Darwin], the complete sanity of his nature, and his wonderful lovableness are, perhaps, the characteristics that will come out most conspicuously in these volumes to those who had not the rare privilege and honour of his personal acquaintance." Emma made sure that copies of *The Life and Letters* were sent to former Down House servants, such as governess Catherine Thorley, butler Joseph Parslow, and gardener Henry Lettington (Loy and Loy 2010).

Emma exerted a great deal of influence over what Francis included in the autobiography associated with *The Life and Letters*. This generated some family friction: Emma and

Henrietta were allied on the side of partial censorship, while Francis and William were in favor of telling it the way Charles wrote it (Moore 1989). The women prevailed. Emma did not want Charles's irreligious views to offend relatives, such as Charles's only living sibling, his sister Caroline; friends like Admiral Bartholomew J. Sulivan; family servants; and the general public. Francis acceded to his mother's "requests" and the omissions are interesting. Several examples are cited by Loy and Loy (2010). In his autobiographical sketch Charles had written, "Nor must we overlook the probability of the constant inculcation in a belief in God on the minds of children producing so strong and perhaps an inherited effect on their brains not yet fully developed, that it would be as difficult for them to throw off their belief in God, as for a monkey to throw off its instinctive fear and hatred of a snake." Emma strongly objected to this statement, and it did not appear in the final version edited by Francis. She also objected to Charles's idea that morality evolved in the same way as physical characteristics—by natural selection. To read Darwin's autobiography as he intended it, one must see the unabridged edition, with original omissions restored, that was produced by his granddaughter (Horace's daughter), Nora Barlow (1958). By then, of course, there would be no relatives, friends, or other Victorians left to be offended.

Darwin biographer Janet Browne (2010) analyzed how Darwin's intellectual development has been portrayed in the approximately 30 biographical studies published since his death in 1882. How Darwin has been depicted has itself evolved as science and society have changed from Victorian norms.

DARWIN'S THOUGHTS ABOUT RELIGION

Darwin spoke for himself about his religious beliefs in the unexpurgated version of his *Autobiography*. A selection of quotes gives a good idea of his views: "Whilst on board the *Beagle* I was quite orthodox.... But I had gradually come...to see that the Old Testament from its manifestly false history of the world...and from its attribution to God of the feelings of a revengeful tyrant, was no more to be trusted than the sacred books of the Hindoos, or the beliefs of any barbarian...[that] the more we know of the fixed laws of nature the more incredible do miracles become...that the Gospels cannot be proved to have been written simultaneously with the events...by such reflections as these...I gradually came to disbelieve in Christianity as a divine revelation." Darwin saw no compelling reason to believe in things that cannot be proved.

When asked, via letter, by an American chemistry professor at Washington and Lee University in Virginia to respond to the Lady Hope deathbed-conversion story, Francis replied: "Neither I nor any member of my family have any knowledge of her [Lady Hope] or of her supposed visits to Down which is quite obviously a work of imagination. He could not have become openly and enthusiastically Christian without the knowledge of his family, and no such change occurred" (Moore 1994). Francis, as the compiler of *The Life and Letters*, was well aware of his father's own words: "In my most extreme fluctuations I have never been an Atheist in the sense of denying the existence of a God. I think that generally (and more and more as I grow older), but not always, that an Agnostic would be the more correct description of my state of mind." Charles, ever attuned to the sensibilities of others, felt that agnostic was a less aggressive word than atheist (Desmond and Moore 1991). Moore (1989) analyzed in great detail why Darwin abandoned Christianity, and Nick Spencer (2009) provided a fascinating modern perspective of Darwin's religious views, or lack thereof.

MORE OF FATHER'S WORK

Francis produced *More Letters of Charles Darwin*, jointly edited with A. C. Seward, in 1903. In 1908, Francis transferred his father's library to the botany school at Cambridge University and indicated that he intended to bequeath it to the university (*Times* [London] 1925). Francis also made available Charles Darwin's two early essays of 1842 and 1844, in which he sketched out the essence of his idea about evolution. These important works, contained within *The Foundations of "The Origin of Species"* (F. Darwin 1909), can be studied by biologists and historians to gain an appreciation for the gestation of Darwin's thinking. The publication date coincided with the Darwin centenary, and Francis received an honorary Sc.D. from Cambridge at this time (Figure 9.5). In 1912 he was honored with the Darwin Medal of the Royal Society, and he was knighted the following year.

FIGURE 9.5 Francis Darwin. National Portrait Gallery, London.

LIKE FATHER, LIKE SON

Frank's friend William Rothenstein, a writer and artist who painted Frank's portrait, wrote: "He has much of his father's directness and simplicity. There was never any doubt about what nor whom he liked and disliked. The sweetest and gentlest of men, he was moved to anger by cruelty, cruelty to animals particularly. Nor could he be indifferent to attacks on his father, which he thought unfair" (Healey 2001).

THIRD MARRIAGE AND FINAL YEARS OF SIR FRANCIS

Frank married the beautiful playwright Florence Henrietta Fisher, the widow of Cambridge law professor Frederic William Maitland, in 1913. They settled down in a large house in Gloucestershire, in the Cotswolds. Florence died in 1920, and Francis returned to Cambridge. He regularly came in to the botany school and examined plants from the botanic garden (Ayres 2008). Francis continued to publish, and his output from 1872 to 1921 numbered some sixty papers. He received honorary degrees from seven universities in Europe and the United Kingdom (Junker 2004). He died at home (10 Madingly Road) in Cambridge on 19 September 1925, at the age of 77. His funeral service was at Christ's College chapel and burial was at the Huntingdon Road cemetery (*Times* [London] 1925).

Francis's son Bernard married Elinor Mary Monsell in 1906. Their son Robert Vere Darwin (1910–1974), a prominent landscape and portrait painter, was married twice but was childless (Goodden 2004).

LEONARD DARWIN (1850–1943)

Charles Darwin was about halfway through his eight-year odyssey into barnacles as Leonard, his fourth son and eighth child, was born on 15 January 1850. The doctor was late in arriving at Down House, so Charles had to administer chloroform to Emma on a saturated cloth. This new anesthetic had only been in use for two years. She became unconscious immediately and remembered nothing about the birth. Charles apparently used too much chloroform, and Emma was out for about 90 minutes. Charles thought this was grand (Burkhardt et al. 1985–, 4: 302–303). Criticism of the drug's use was silenced when Queen Victoria gave birth to Leopold, her eighth child, under its influence in April 1853 (Healey 2001).

LENNY THE WISE GUY

The name Leonard was chosen in honor of Leonard Jenyns, a naturalist and friend from Cambridge days whom Charles enlisted to write the fish section of *The Zoology of the Voyage of the H.M.S. Beagle.* Jenyns was John Henslow's brother-in-law, and Henslow also had a son named Leonard (Burkhardt et al. 1985–, 4: 303–304). The new arrival quickly became known as "Pouter," then "Lenny," and later as "Leo."

Francis Darwin told the story, quoted by Browne (1995), about Lenny jumping on the sofa, which was not allowed in Down House. Their father caught him doing this and said, "Oh Lenny Lenny it is against all rules." To which Lenny replied,

FIGURE 10.1 Emma Darwin (at about 45 years old) and three-year-old Leonard Darwin, demonstrating his nickname "Pouter." A pouter was also a breed of long-legged domestic pigeons, characterized by their habit of puffing out the distensible crop. Photograph by Maull and Fox, circa 1853. Darwin Museum, Down House.

"Then I think you'd better go out of the room." Lenny became known for his jokes and wisecracks. He was a typical Darwin child in his zeal for beetle collecting (Figure 10.1). He was also a budding philatelist and took great delight in the stamps that Asa Gray sent to him from America (Loy and Loy 2010).

SCARLET FEVER

Both Lenny and Franky developed a fever in 1855, and there were other illnesses to come for Lenny. By the time he was

11 years old, his father felt that Lenny was a bit backwards in learning his lessons because of his illnesses. He was tutored by the local vicar, George Varenne Reed, before being enrolled in Clapham School at the age of 12. This grammar school was close enough to allow the boys to return to Down House on weekends. On 12 June 1862, Lenny was sent home from school because he had scarlet fever (Browne 2002). Several weeks of grave concern for his life permeated Down House. He relapsed in mid-August and Emma also developed scarlet fever. The end result of all this was a very ill Charles Darwin, who worried himself sick about the health of his family.

DARWIN-MENDEL VISIT THAT NEVER HAPPENED

Leonard's niece, Margaret Elizabeth (George's daughter), wrote a lovely memoir about her uncle for the *Economic Journal*. In that piece she also recounted a very interesting scenario, pregnant with implications for the history of biology, which I will use for a related digression (M. Keynes 1943). The story is that Gregor Mendel, the Austrian friar who, by experimenting with pea plants, had formulated the first two laws of heredity (segregation, and independent assortment), was in Downe in 1862 and could have visited Charles Darwin. Because of Lenny's near-death experience from scarlet fever, however, this meeting did not take place. The mind boggles at what could have resulted had the two ever met!

Mendel published his results in 1866 in an obscure German journal, but this paper was not widely known until Hugo de Vries (Dutch), Carl Correns (German), and Erich von Tschermak (Austrian) more or less simultaneously rediscovered it in 1900. Mawer (2006) unraveled the behind-the-scenes maneuvering this rediscovery produced in the scientific literature. William

Bateson (English) had Mendel's paper translated into English and published. Bateson became a staunch supporter of Mendel, coined the word genetics for the study of inheritance, and eventually demonstrated that the same Mendelian hereditary patterns in plants also occur in animals.

Here is what Leonard wrote 80 years later, when contacted by a German who told him he believed Mendel had been to Down in 1862 (M. Keynes 1943):

> I have told him that I am fairly certain that the interview never took place. I am the only person who can now fully realize what an event the appearance of a German Catholic priest at Down would then have been. There is also another reason in regard to which I oddly enough play the chief role. At as nearly as possible when Mendel was in England, I was recovering from a long and very dangerous illness, and my mother was suffering from scarlet fever caught from me. If I prevented my father from meeting Mendel, do you not think that I even now ought to be hung, drawn and quartered? Moreover I think people now forget the prejudices of those days: the Catholic clergy are much more under the thumb of their superiors than are the protestants, and I only remember 3 clergy coming to Down House, 2 local parsons, and [Charles] Kingsley. I do not believe Mendel would have thought it wise to get the reputation of having seen my father.

Mendel was influenced by Darwin's research and, in fact, even owned a copy of *On the Origin of Species*, but there is no evidence that Darwin was aware of Mendel's experiments (Mawer 2006). Such knowledge would have enabled Darwin to abandon his mistaken notion of pangenesis and gemmules, which he postulated in *Variation of Plants and Animals under Domestication* in 1868. Darwin's system of inheritance involved the shedding of gemmules by cells and tissues throughout the body. The particles reached the gonads via the bloodstream and became part of the sex cells, which were passed on to offspring. One can only wonder what Darwin would have said and done if he had been exposed to Mendelian genetics.

The spurious story, recounting that Darwin had been sent one of the 40 reprints (offprints) Mendel ordered of his paper in the 1866 *Proceedings of the Brünn Society of Natural History* but had not read it, nonetheless contains a wisp of truth, as explained by Mawer (2006). Mendel's paper, "Experiments in Plant Hybridization" ("Versuche über Pflanzen-Hybriden"), was cited in an important book by W. O. Focke, published in 1881, entitled *Die Pflanzen-Mischlinge* (*Plant Hybridization*). Darwin owned a copy of this book, which is still in his library, but the pages dealing with Mendel's work are uncut (the bound, folded pages are unopened), meaning they could not have been read.

Of the 40 reprints of Mendel's pea-plant paper, only 7 have been located (Mawer 2006). One of these is in the Indiana University library, and was acquired by Dr. Charles Davenport about 1898. Davenport was a zoologist and one of the first US scientists to utilize Mendelian genetics (Berra et al. 2010b). He established the Eugenics Record Office (ERO) of the Department of Genetics at the Carnegie Institute in Washington, D.C. The ERO was located in Cold Spring Harbor on Long Island in New York State. Davenport hired Harry H. Laughlin to be superintendent of the ERO in 1910. This became the foremost center for eugenics in the United States. One of its missions was to collect the pedigrees of prominent families. Laughlin eventually produced a large pedigree of the Galton-Darwin-Wedgwood families (front endpapers) and was well acquainted with Leonard Darwin, who had been president of the Eugenics Society.

LIFE AT DOWN HOUSE

Browne (2002) related a story about Leo's 1862 return home from Clapham, during his first term there. *The Origin* had been the topic of conversation among his schoolmates, and Leo wanted to read it: "I remember my father entering the drawing room at Down, apparently seeking for someone, when I, then a schoolboy, was sitting on the sofa with the *Origin of Species* in my hands. He looked over my shoulder and said: 'I bet you half a crown that you do not get to the end of that book.' " Charles "won his bet but never got his money."

Down House was open to the public as a national memorial and museum on 7 June 1929, initially under the care of the British Association for the Advancement of Science and, later, the Royal College of Surgeons (Reeve 2009). This occasion caused Leonard, the only surviving child of Charles and Emma, to reminisce about his childhood at Down (L. Darwin 1929). His memories are a gold mine of information about his father and life at Down House. Leonard described how he climbed a large holly tree one winter in his Sunday clothes. He received such a scolding from his father as "to wipe out for the rest of my life every desire to climb trees in my Sunday best."

Leonard thought that Charles overstated his lack of aesthetic emotions in his *Autobiography*, because he recalled his father declaring "that if he had to live his life over again he would make it a rule to let no day pass without reading a few lines of poetry." Leonard further added, "It seemed to all of us onlookers that his appreciation of natural scenery remained quite undimmed to the end of his life." Leonard explained that his "father's life cannot be understood without reference to what he suffered." When his father was asked by one of the children if he could not go away from home and rest, Charles "replied that the truth was that he was *never* quite comfortable except when utterly absorbed in his writing. He evidently dreaded idleness as robbing him of his one anodyne, work" (L. Darwin 1929).

LEONARD ON HIS FATHER'S ILLNESS

Much has been written about Charles Darwin's ailments (Colp 2008; Berra et al. 2010b; and references cited therein). Here is what Leonard had to say on the subject (L. Darwin 1929):

> No doctor seemed to know from what he was suffering, and in his own opinion it was not the result of sea-sickness on his long voyage, severe though that trouble had been. Though it is very rash for a layman to speak on such subjects, yet I cannot refrain from recording my belief that it was pyorrhea [peridontitis], or some other form of auto-poisoning, and that any excitement made the poison flow more freely. It is in any case a fact that for many years an hour's interesting conversation in the afternoon with a visitor would bring on several hours' vomiting during the succeeding night, whilst he was hardly ever without some symptoms of indigestion.

Leonard goes on to marvel at the mass of work his father accomplished in spite of his ill health: "It was only done by declining to undertake extraneous duties—and here his bad health was a real help—and by never wasting a single minute of his short day's work." Nevertheless, Leonard wrote, "my father spoke of his life as a happy one, and this was certainly true, though it was greatly marred by very long periods of discomfort and suffering, which mercifully got decidedly less frequent towards the end of his life."

The penultimate in a long litany of diagnoses-at-a-distance for what ailed Charles Darwin (Crohn's disease, systemic lactose intolerance, Chagas disease, arsenic poisoning, allergies, psychosomatic disorders, etc.) is chronic vomiting syndrome (Hayman 2009a, 2009b). The author, a Melbourne University pathologist, recently informed me that his preferred diagnosis, based on circumstantial evidence, is now an A to G point mutation at nucleotide 3243 in a mitochondrial ring chromosome, and that this explains all of Darwin's symptoms (John Hayman, personal communication, 22 August 2012; Hayman 2013).

ARMY CAREER

In spite of his slow start, Lenny, now 18 years old, did well at Clapham and placed second on the entrance exam to attend the Royal Military Academy at Woolwich, where he pursued an engineering education. Charles was very proud of his son's late-blooming accomplishments. Leonard's special interest was photography, which was very useful in survey operations. He was commissioned in the Royal Engineers in 1870. In his usual self-effacing way, Leonard claimed that he joined the army because he thought himself the dimmest of the Darwin sons (M. Keynes 1943), but he told his niece Gwen that it was "because he was afraid of being afraid" (Raverat 1952). Leonard had a 20-year army career.

The Franco-Prussian War began on 19 July 1870, encouraged by the ruler of France (Napoleon III) and the prime minister of Prussia (Otto von Bismarck). The war was ostensibly over succession to the Spanish throne. Casualties were very high. The French were overwhelmed by the German war machine, Napoleon III surrendered, and Paris capitulated on 28 January 1871. Bismarck used the victory to complete the unification of Germany. England remained neutral. Leonard (as well as Charles and Emma), however, favored the Prussians, but most of Leonard's fellow soldiers wanted the French to win, so there could be an excuse for a war with France, Britain's traditional enemy (H. Litchfield 1915, 2: 198–199).

After the war but prior to the publication of *The Expression of Emotions in Man and Animals*, Charles turned over the proof correction to Leonard and Henrietta. They made extensive changes to this pioneering work that used photographs to illustrate the text, and all were gratified to see it published on 26 November 1872. This work includes 32 photographs and 21 woodcuts. Because photographs taken during that period required long exposure times, animal expressions are represented in the woodcuts, and those of humans in the photographs (Voss 2010). Darwin was attempting to illustrate that human expressions are a physiological response to stimuli, and thus are vestiges of the same response seen in other animals.

In 1874 Leonard and a contingent of Royal Engineers were posted to New Zealand, where Leonard was to observe and photograph a transit of Venus on 9 December. Unfortunately the sky was cloudy and the event could not be seen. He returned home by steamer via San Francisco and then took a train to Washington, Philadelphia, New York City, and Boston. In Boston he visited Asa Gray, his stamp supplier from childhood. The Atlantic crossing was on the S.S. *Abyssinia* of the Cunard Line. Leonard then served a two-year posting in Malta. While on leave in March 1876, Leonard and George toured around Italy (Loy and Loy 2010). Leonard taught chemistry and photography at the school of military engineering at Chatham from 1877 to 1882.

PHOTOGRAPHY

Charles had encouraged photography as a hobby for both William and Leonard in their childhoods. The chemicals were expensive, however, and the outcomes were often unreliable. Leonard eventually produced some well-known photographs of his father (Figure 10.2). The most famous is the photograph of Charles in a wicker chair on the veranda of Down House circa 1878 (Berra et al. 2010b). The veranda was constructed in 1872 and faced the garden. It was the family's favorite place to sit. Photography was in its infancy in the nineteenth century and exposure times were long. This may explain why very few

FIGURE 10.2 Charles Darwin (circa 69 years old), in a wicker chair on the veranda of Down House. Photographed by Leonard Darwin, about 1878. Darwin Museum, Down House.

smiles were recorded from that period, as it was too long to hold one's facial muscles in a smile. Posing for a photograph was not all that different from sitting for a painting. In 1878 Leonard introduced the Darwin household to the telephone, probably a military field version, but it was less than a rousing success (Browne 2002).

MARRIAGE TO BEE

Leo and his fiancée, Elizabeth ("Bee") Frances Fraser, visited Charles at Down House on 23 March 1882, only a month before Charles's death. She was the sister of one of Leonard's fellow officers. When Leo's father died in April, he hurried home to join his other brothers. Newly promoted Captain Leonard Darwin and Bee were married on 11 July 1882. He was then ordered to Jimbour (near Brisbane in Queensland, Australia) to observe another transit of Venus that was supposed to occur on 10 December 1882. His new wife accompanied him, and they had their honeymoon during the trip to Australia. While sailing there, Leonard learned that he had earned a first in his Staff College examination (Loy and Loy 2010). Alas, it was again too cloudy to carry out his mission. Leonard wrote, "There are few people who have been twice round the world to see a thing without seeing it" (M. Keynes 1943).

On the way home, Leonard and Bee stopped in Singapore. Leonard used telegraphic signals to calculate the longitudinal difference between the Singapore telegraph station and Port Darwin, the northern Australian town named for his father (Loy and Loy 2010). The couple returned to England at the end of April 1883, and he resumed teaching chemistry at the Staff College.

MAJOR LEONARD

Leonard worked in the Intelligence Service of the War Office in London from 1885 to 1890 and was promoted to major in 1889. He was assigned to the African desk, where he worked in the topographical and colonial sections, which dealt with West African colonies. He was sent to Paris to delimit various colonial borders. In 1886 he was seconded to the Royal Society and sent on another astronomical expedition, this time to Grenada in the West Indies (Loy and Loy 2010). His wife went with him, and they arrived on 12 August. His assignment was to photograph a total eclipse of the sun, and the weather finally cooperated. His observations on photographing the corona and solar prominences with a prismatic camera were published in the *Philosophical Transactions of the Royal Society* (L. Darwin et al. 1889).

RETIRED FROM THE ARMY

Leonard resigned his military commission in 1890. Doctors had recommended a long sea voyage for the health of his wife, so Leonard and Bee sailed for New York City, crossed to California, and then went on to Japan, China, and Egypt on a six-month circumnavigation of the globe (Loy and Loy 2010). Leonard was the most widely traveled of all the Darwin children. Treasures acquired on this trip furnished and filled his and Bee's home.

On his return to England, Leonard joined the Royal Geographical Society and was elected to its council in 1890. He became its president in 1908 and served until 1911. He wrote a few pieces for the *Dictionary of National Biography*, including a sketch of Thomas Wedgwood (L. Darwin 1900), but he found this too dull and without financial reward (Loy and Loy 2010).

DARWIN, NORTHERN TERRITORY, AUSTRALIA

Since I am privileged to be a university professorial fellow at Charles Darwin University and research associate at the Museums and Art Galleries of the Northern Territory, both located in Darwin, Northern Territory, Australia, I would be remiss if I did not include the following story from Margaret (M. Keynes 1943). When questioned about the origin of the name Port Darwin, Leonard wrote in 1941:

> If you look at the chart in your edition of the *Voyage of the* Beagle, you will see that she [the H.M.S. *Beagle*] never could have been within a thousand miles of Port Darwin when my father was on board, so the name could not have been given then. My memory & not too clear—is that my father never *heard* how the name was

> given; but that he knew that an officer from the *Beagle*, I presume [John] Stokes, had subsequently surveyed the northern coast of Australia, and he felt sure that he had then given it that name in memory of their voyage together. I do not think it strange that he [Charles] never heard of this event at the time. When I was president of the R. G. S. two mountains were called after me.... [One] was [named] by Scott (?) in the Antarctic. He probably said: "Call it after the President of the R. G. S." and he did so and thought no more of the matter, having to name places every day. I don't know if that Mount Darwin still appears or not. As to Port Darwin, when named it was a dreary desolate bay, with behind it the country since called the Never-Never land, an inhospitable region. Stokes might have felt that giving it the name of Darwin was no great compliment.

JOHN LORT STOKES

On 27 April 1882 the *Times* [London] printed an account of Charles Darwin's funeral. Next to this article was a letter to the editor by Admiral John Lort Stokes, reminiscing about his and Charles's shipboard experiences in 1831–1836. He commented on Darwin's seasickness and quoted Charles: "Old fellow I must take the horizontal for it." Stokes elaborated, "It was distressing to witness this early sacrifice of Mr. Darwin's health, who ever afterwards seriously felt the ill-effects of the *Beagle*'s voyage." Stokes was a 19-year-old mate and assistant surveyor on the second voyage of the H.M.S. *Beagle*, with Charles aboard, when she sailed in 1831. He was also Charles's roommate in the tiny poop cabin, along with midshipman Philip Gidley King, for the nearly five-year voyage. It was on the third voyage of the *Beagle* in 1839 that naval surveyor Stokes named Darwin Harbour for his friend. Stokes joined the navy at the age of 13, and spent 18 years on all three voyages of the H.M.S. *Beagle* (Powell 2009).

He was in command of the *Beagle* by the end of the third voyage (Freeman 1978).

POLITICS

Leonard had three careers: (1) military, (2) politics and economics, and (3) eugenics. After leaving the army he served on the London County Council as a Moderate. He ran for election to Parliament as the Liberal-Unionist member for the Lichfield division of Staffordshire. With campaign support from his wife, he won the general election in July 1892, although only by four votes. His mother enthusiastically supported his candidacy and threw a party to celebrate his election (Healey 2001). He served as a member of Parliament from 1892 to 1895, but was defeated for reelection in 1895. Leonard was not cut out for the realities of politics. As Sir Arthur Keith remarked, Leonard considered politics "as the art of applying science to the problems of government," and Leonard had a penchant for seeing both sides of every question (M. Keynes 1943).

Leonard also gave of himself to public service in other ways. In 1892 he became a member of the council of Bedford College for Women. His Uncle Erasmus (Charles's brother) played a leading role in its founding in 1849. Leonard worked for the college for 33 years. From 1913 to 1920 he was chairman of Bedford College of London University. During World War I he served as chairman of the Professional Classes Relief Council from 1914 to 1918, which reflected his view that eugenically desirable qualities were segregated in social classes. This commission helped a socially selected group through economic difficulties. It was an early experiment in social welfare, related to social biology, that was designed to "encourage parenthood and to give help to those carrying good stock through the tribulations of the war period" (*Eugenics Review* 1943).

ECONOMICS

Leonard's experiences in government led to his involvement in economics. He wrote the very influential textbook *Bimetallism* (1897) (Figure 10.3), which was highly praised by economist J. Keynes (1943). John Maynard Keynes (1883–1946) was one of the world's most influential economists, and his theories remain in use. Keynesian economic policy, which calls for government spending during a recession, is still part of today's economic debates. As the title, *Bimetallism*, implies, Leonard's subject matter dealt with the price of gold and silver. It argued that the government ought to define the value of its monetary unit in terms of both metals, thus establishing a fixed rate of exchange between them. Leonard was a moderate adherent to the policy of allowing the market to determine the ratio between gold and silver prices, instead of favoring a high gold-price for silver, which could raise actual gold prices and increase inflation. He wanted to stabilize the exchange rates between countries that pegged their currency to either gold or silver. As Miss Prism says in *The Importance of Being Earnest*, "When one has thoroughly mastered the principles of Bimetallism one has the right to lead an introspective life. Hardly before. I must beg you to return to your Political Economy."

The array of eclectic topics Leonard tackled is really quite astonishing. For example, he testified before a parliamentary committee about Indian currency (L. Darwin 1899). In 1903 he published an extensive study entitled *Municipal Trade: The Advantages and Disadvantages Resulting from the Substitution of Representative Bodies for Private Proprietors in the Management of Industrial Undertakings*. In it, he concluded that private management was more efficient than public management. This led to a series of lectures at Harvard University that became another book, *Municipal Ownership* (L. Darwin 1907).

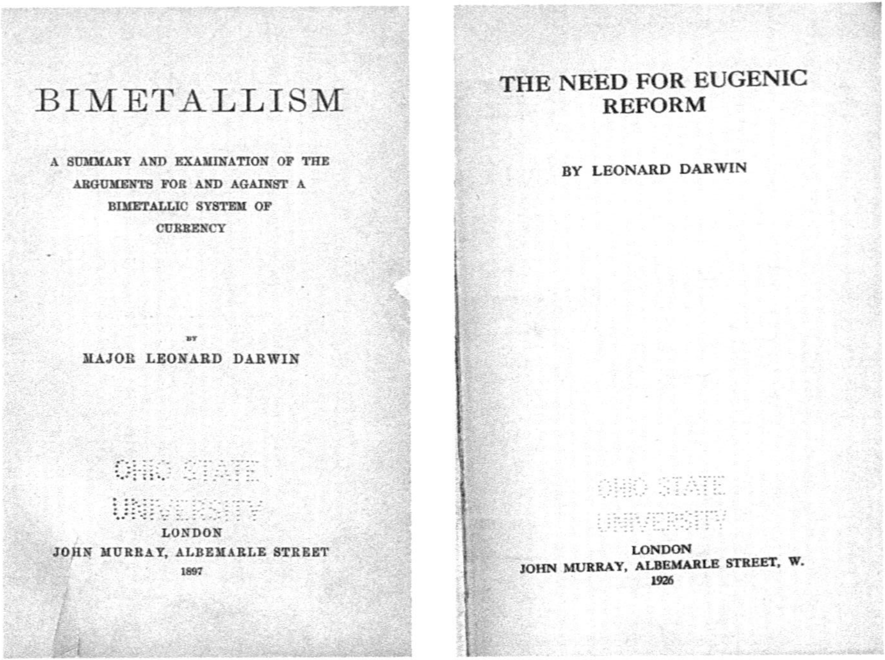

BIMETALLISM

A SUMMARY AND EXAMINATION OF THE ARGUMENTS FOR AND AGAINST A BIMETALLIC SYSTEM OF CURRENCY

BY

MAJOR LEONARD DARWIN

LONDON
JOHN MURRAY, ALBEMARLE STREET
1897

THE NEED FOR EUGENIC REFORM

BY LEONARD DARWIN

LONDON
JOHN MURRAY, ALBEMARLE STREET, W.
1926

FIGURE 10.3 *Left*: The title page of the first edition of *Bimetallism*, published in 1897. *Right*: The title page of the first edition of *The Need for Eugenic Reform*, published in 1926.

MARRIAGE TO MILDRED

Leonard and Bee had no children. She died in 1898, after a long illness. He married Charlotte Mildred Massingberd, his second cousin, in 1900. She was 18 years his junior. The marriage prospered for 40 years, but again there were no children. She was known as Mildred within the family.

EUGENICS

Leonard's studies led him to consider the relationship between economics and heredity, thus ushering in his third career in eugenics. Leonard was 61 years old when he turned to eugenics, which was to become his major interest in life. The eugenics movement was founded by Sir Francis Galton, Charles Darwin's half first cousin and thereby Leonard's half first cousin once removed. Galton coined the term eugenics in 1883, and he founded the Eugenics Education Society in 1908 to encourage the study of human heredity. The word "education" was later dropped from the Eugenics Society's name. Many in the society (including Leonard) felt that its objective should be limited "to the creation of a public opinion favorable to the promotion of fertility among those who could enrich the biological endowment of posterity [positive eugenics] and to its restriction, by sterilization or other means, among those whose contribution to posterity could be spared [negative eugenics]" (*Eugenics Review* 1943). After Galton's death in 1911, Leonard, against his strong resistance, was persuaded to become president of the society, a post he held until 1928. This enabled him to combine his interest in human affairs with science (M. Keynes 1943). In 1912, Cambridge University bestowed an honorary degree of Doctor of Science on Leonard.

As he did in politics and economics, Leonard steered a middle course and maintained the equality of environment and heredity. In a letter to the editor of the *New York Times* (21 December 1912, p. 12), Leonard, as president of the Eugenics Society, spelled out his view of the purposes of eugenics: "We desire therefore greatly to increase the sense of responsibility in connection with all matters pertaining to human parenthood, to spread abroad knowledge of the laws of heredity as far as now known, and to encourage further research in that domain of science.... [W]e do not advocate any interference whatever with the free selection of normal mates in marriage.... There will no doubt always remain a class quite outside the pale of all moral influence, and of these there will be a small proportion who, if they become parents, are certain to pass on some grievous mental or bodily defect to a considerable proportion of their progeny. Here and here only must the law step in. As to whether surgical sterilization should ever be enforced on such persons we have still an open mind, but certainly not till further information on this subject in available." He further stated that "sufficient control must be maintained over them [those described above] to prevent them from breeding." He ended his letter by writing that the poor laws should be administered "so as not to encourage reproduction on the part of degenerate paupers."

Leonard was president during the First International Eugenics Congress in London in 1912. In 1921 he spoke at the Second International Eugenics Congress in New York City. He published *The Need for Eugenic Reform* in 1926 (Figure 10.3), which was followed by *What is Eugenics?* It was in 1932, at the Third International Eugenic Congress in New York City, where a large Galton-Darwin-Wedgwood pedigree was exhibited (Berra et al. 2010b). The negative eugenics advocated by Leonard is shocking to today's sensibilities, but it was a product of the times.

Raverat (1952) described fierce arguments that she had with her uncle about eugenics. She was shocked when Uncle Lenny considered a money standard for deciding who should be encouraged to breed. He reportedly said, "A man who can earn and keep money shows that he has the qualities essential to survival." Gwen replied that money means little to artists, philosophers, and other creative people, and she did not want to see those qualities bred out of humans. Uncle Lenny and Aunt Mildred had little use for this argument. On the other hand, as Raverat pointed out, Leonard intervened when the local government wanted to lock up an old man who lived wild in the woods. Leonard knew that would kill the man and argued that he should be allowed to roam the forest freely.

MENTORING RONALD A. FISHER

Perhaps Leonard's greatest contribution to science was his encouragement and support for a young scientist, Ronald A. Fisher, a population geneticist and statistician. They carried on a 20-year correspondence from 1915 to 1935, often writing every few days. Many of those letters are included in Bennett's (1983) review of their relationship. It was Leonard who suggested the topic of Fisher's important paper, "The Correlation between Relatives on the Supposition of Mendelian Inheritance" (R. Fisher 1918). Leonard seemed to be fulfilling the role of father figure and major professor to the brilliant, much younger Fisher (Bennett 1983). Leonard personally defrayed some of Fisher's publication costs (Edwards 2004). The two became fast friends, despite the 40-year age difference, and Leonard was a major influence on Fisher's life and research into biometry, heredity, and selection (Bennett 1983). For Fisher, Leonard Darwin was a living link to Charles Darwin and Francis Galton.

With his characteristic self-effacement, Leonard referred to himself as "muddleheaded" and "stupid about mathematical things." He wrote that he liked receiving Fisher's letters because they always made him think. Bennett (1983) wrote: "He once summed up his feelings on receiving a letter from Fisher as 'somewhat like that of a pig genuinely admiring a necklace of pearls, but not knowing quite how to put it on and feeling sure that he had not deserved such a present.' " Between 1915 and 1935, Fisher published about 200 papers in the *Eugenics Review*, a quarterly journal of the Eugenics Society of which Leonard was president.

Fisher became part of the neo-Darwinian modern evolutionary synthesis that united natural selection with genetics. His classic work, *The Genetical Theory of Natural Selection* (R. Fisher 1930), is dedicated "to Major Leonard Darwin in gratitude for the encouragement, given to the author, during the last fifteen years, by discussing many of the problems dealt with in this book." Leonard had pushed Fisher "to write a great work on the mathematics of evolution," and he certainly got it. This book is required reading in most population genetics and evolutionary biology classes. One of the references Fisher cited was Leonard's book, *The Need for Eugenic Reform* (L. Darwin 1926). About this book Fisher wrote: "It is one of the difficulties of the subject that eugenics exercises a potent attraction for cranks of various kinds. All the more valuable is the sober judgment, detached reasoning and well-weighed earnestness of this really great book" (*Eugenics Review* 1943).

RETIREMENT

In 1921 Leonard and Mildred retired to their country home, Cripps Corner, located on the outskirts of Ashdown Forest

FIGURE 10.4 Leonard and Mildred Darwin, photographed in 1923, Uppsala, Sweden. Cropped from a photo in Berra et al. 2010b.

(Figure 10.4). They lived in this house in relative isolation, as it was a two-hour train ride from London. Gwen Raverat (1952), Leonard's niece and Margaret Keynes's sister, described Aunt Mildred as a fanatical teetotaler, but she allowed Leonard to have his shot of whiskey every night as "medicine," provided he drank it in one gulp. She would not allow electric lights or a telephone, but she did grant Leonard the privilege of owning a motorcar. She was an ardent feminist, but thought women should treat men kindly and not expect too much from them, "because they are such helpless things, Poor Lambs." Apparently the entire Darwin extended family took great delight in Mildred's idiosyncrasies.

World War II came to Cripps Corner in 1940, when part of the Battle of Britain was fought over their heads (M. Keynes 1943). Mildred died in December 1940 of an unspecified illness. Leonard remained alert but became increasingly more feeble. After being sick for three days, he died peacefully from bronchial pneumonia on 26 March 1943, at the age of 93. His obituary appeared the next day, on 27 March (*Times* [London] 1943). He holds the record for longevity among the Darwin children. He is buried in Forest Row cemetery.

In 1943 the Eugenics Society printed a group of comments from its members upon Leonard's death. Most pointed out that he was utterly devoid of personal ambition and was very modest, intellectually honest, and unfailingly courteous to all.

In a letter to Margaret Keynes, dated 11 January 1944, Ronald Fisher wrote that Leonard "was surely the kindest and wisest man I ever knew." Arthur Keith (1943) remarked in Leonard's obituary in *Nature* that in physical appearance and in his attitude toward life, Leonard "bore a closer resemblance to his father than did any of his brothers." As reported by Raverat (1952), Leonard's nephew Bernard Darwin (Frank's son) wrote this poem about his Uncle Lenny:

Serenely kind and humbly wise,
Whom each may tell the thing that's hidden
And always ready to advise
And ne'er to give advice unbidden.

HORACE DARWIN (1851–1928)

Spring 1851 was an extraordinarily trying time for the Darwin family. Annie died on 23 April at Malvern, leaving Charles devastated. Emma, 43 years old and deeply depressed by Annie's loss, was eight months pregnant. She fought through the sadness, and, aided by chloroform, Horace was safely delivered on 13 May. Charles was still working on his barnacles. The house was full of curious and active children. Emma had given birth 9 times in 12 years. Surely she supposed Horace would be the last. No one can say that the Darwins did not have an active sex life!

ENTHUSIASM FOR BEETLES AND MACHINES

In July 1858 the entire family, including seven-year-old Horace, enjoyed a month-long vacation on the Isle of Wight. Charles was pleased that Horace showed the usual Darwin enthusiasm for beetle collecting. A note on the beetles of Downe appeared in the *Entomologist's Weekly Intelligencer* in 1859, bearing the names of Francis, Leonard, and Horace as its authors (Freeman 1982). Since they were 10, 8, and 7 years old, respectively, their father must have organized this, probably to encourage their natural-history investigations.

Just as William's and Leonard's interest in photography may conjure up the spirit of Thomas Wedgwood, Horace had an interest in machinery that would make his great-grandfather, Josiah Wedgwood I, proud. The innovative senior Wedgwood designed and built a pyrometer in 1782 to measure the temperature of the furnaces in which he fired his clay pottery. Perhaps this was a prescient influence behind Charles and Emma's son becoming a world-class instrument maker. They encouraged his passion by converting the schoolroom at Down House into a workroom for Horace, complete with a lathe and other tools (Cattermole and Wolfe 1987).

A VERY PERCEPTIVE BOY

Leonard Darwin dictated a brief sketch of Horace's life to his wife Mildred on 12 June 1932, and she transcribed it in longhand. This mini-biography was reproduced in Cattermole and Wolfe (1987) and contains an interesting look into the intellectual life of 11-year-old Horace (Figure 11.1). Charles Darwin, in a letter to Lord Avebury, wrote: "Horace said to me yesterday, 'If everyone would kill adders they would come to sting less'. I answered, 'Of course they would, for there would be fewer'. He replied indignantly, 'I did not mean that; but the timid adders which run away would be saved, and in time they would never sting at all'. Natural selection of cowards!" (F. Darwin and Seward 1903, 1: 204). I can only wish that today's students would grasp natural selection so intuitively.

ONE MORE SICKLY DARWIN CHILD

By age 11, Horace developed "the shakes" and had become another ailing child for his parents to worry about. Multiple doctors at that time variously diagnosed this condition as concussion, digestive irritation, or roundworms (Loy and Loy 2010). Charles wrote to his first son, William, on 14 February 1862 (Burkhardt et al. 1985–, 10: 80): "We have been very miserable,

FIGURE 11.1 Horace Darwin. Cambridge University Library.

& I keep in a state of almost constant fear, about poor dear little Skimp, who has oddest attacks, many times a day, of shuddering & gasping & hysterical sobbing, semi-convulsive movements, with much distress of feeling. . . . We shall have no peace in life till the poor dear sweet little man gets better." Charles, as usual, thought this was a peculiar form of inheritance from his own poor constitution (Burkhardt et al. 1985–, 10: 92). Several doctors were consulted, to no avail.

The fits may have been a combination of the Darwin hypochondria and a plea for attention from the children's young German governess, Camilla Ludwig, of whom Horace was extraordinarily fond (Browne 2002). The insightful Emma sent Camilla back to Germany to visit her mother and took Horace to visit William in Southampton. This seemed to break the spell of Camilla, and Horace recovered his equilibrium. Charles was appreciative of Camilla, who helped him translate German scientific papers. She eventually returned to Down House. In 1863 Horace, still a sickly child, joined his father in taking the waters at Malvern.

Horace was tutored by the local clergyman—as were his brothers George, Francis, and Leonard—and then sent to Clapham School, where he was looked after by 18-year-old Lenny, one year his senior. Like a typical Darwin, Horace had his share of illnesses and returned home for nursing from time to time. In spite of his ill health, which improved after he was 12 years old, Horace loved high jumping and riding horses, and he was considered physically strong by his brother Leonard (Cattermole and Wolfe 1987).

HOLIDAY ON THE ISLE OF WIGHT

In July 1868 Charles was feeling unwell, and he, Emma, and 17-year-old Horace took a vacation on the Isle of Wight. The Darwins rented the home of photographer Julia Margaret Cameron at Freshwater Bay. Charles's brother Erasmus and Joseph Dalton Hooker also made the trip to Freshwater Bay, not wanting to miss the visitations with Alfred Tennyson and Henry Wadsworth Longfellow that Mrs. Cameron could arrange (Browne 2002). Longfellow was distantly and convolutedly related to the Darwins by marriage, in that he was the brother-in-law of Fanny Wedgwood's (Hensleigh's wife) sister-in-law, Mary Mackintosh, an American (Berra et al. 2010a, 2010b; Loy and Loy 2010). Julia Cameron managed to photograph Charles (Figure 11.2) and all the available Darwins except Emma. Charles inscribed his compliments on the right-profile picture she took of him: "I like this photograph very much better than

FIGURE 11.2 Charles Darwin, at 59 years old. Photographed by Julia Margaret Cameron at Dumbola Lodge, Freshwater, Isle of Wight, in July 1868.

any other which has been taken of me" (F. Darwin 1887, 3: 92; Prodger 2009).

Another portrait taken during this session shows Darwin in left profile (see Burkhardt et al. 1985–, 16: facing 630). An engraving based on this photograph appeared in 2000 on the £10 English banknote—with Charles Darwin replacing Charles Dickens (Voss 2010). Darwin, however, considered this particular photograph to be "heavy & unclear" (Burkhardt et al. 1985–, 17: 515). Another Cameron photograph of Darwin was signed "Ch. Darwin March 7th 1874" and used as a calling card (Prodger 2009; Voss 2010).

Mrs. Cameron often used young girls as models (Voss 2010), but she felt that no woman between the ages of 18 and 70 should be photographed (Healey 2001). She also had taken

FIGURE 11.3 Seventeen-year-old Horace Darwin, photographed by Julia Margaret Cameron, on the Isle of Wight, in 1868.

a special interest in the delicate young Horace (Figure 11.3), who seemed to have an attraction for and to older women (Healey 2001). The Darwins' visit lasted nearly six weeks.

DARWIN'S HORSE ACCIDENT

The following year, on 9 April 1869, Charles's old horse, Tommy, stumbled, and Charles fell. Tommy fell on Charles, whose back and leg were injured by the blow (Healey 2001). Although he had previously been a daring horseman during the H.M.S. *Beagle* voyage and had ridden with gauchos in South America, 60-year-old Charles vowed that this was the end of his riding days. Fortunately, his injured leg recovered in a week or so. During that time he couldn't walk, and Horace pulled him on a trolley to the greenhouse, so Charles could make observations on his ongoing botanical experiments (Browne 2002).

ANOTHER DARWIN AT CAMBRIDGE

Horace eventually joined Francis at Trinity College in Cambridge in 1868, but Horace took six years to earn his degree, due to repeated illnesses. Upon learning that Horace had passed his set of Cambridge exams known as the "Little Go," Charles wrote his son a letter, dated 15 December 1871. It showed Charles's pride in his son's accomplishment as well as a bit of Charles's character, and the letter was reproduced in the *Times* [London] 1928 obituary for Horace on 24 September 1928:

> My Dear Horace,
>
> We are so rejoiced, for we have just had a card from that good George in Cambridge saying that you are all right and safe through the accursed "Little Go". I am so glad, and now you can follow the bent of your talents and work as hard at mathematics and science as your health will permit. I have been speculating last night what makes a man a discoverer of undiscovered things; and a most perplexing problem it is. Many men who are very clever—much cleverer than the discoverers—never originate anything. As far as I can conjecture, the art consists in habitually searching for the causes and meaning of everything which occurs. This implies sharp observations, and requires as much knowledge as possible of the subject investigated. But why I write all this now I hardly know—except out of the fullness of my heart; for I do rejoice heartily that you have passed this Charybdis.
>
> Your affectionate father,
>
> C. Darwin.

ENGINEERING APPRENTICE

Horace received his B.A. in mathematics in 1874 and then undertook a three-year apprenticeship with Easton and Anderson, an engineering firm in Erith, Kent (Glazebrook 2004). This aroused his interest in scientific instruments, and, while there, he built a device for measuring a plant's response to stimulus (Glazebrook 1928). This "klinostat" was the first instrument he designed, and Francis Darwin used it for recording the rate of plant growth. Horace also learned pattern-making and foundry practices. He returned to Cambridge at the end of his apprenticeship and resided there for the rest of his life. Horace earned a living as an engineering consultant on land reclamation and drainage projects. His university contacts frequently asked him for help in designing and constructing various kinds of instruments and assorted apparatuses needed for precision measurements and the recording of events. Horace joined civil and mechanical engineering societies in 1877–1878.

WORM STONE

Horace also invented the "worm stone" used by his father to calculate the rate at which stones on the surface were buried by the excavation activities of earthworms beneath them. This device involved a 23-kilogram (50.6 pound) flat stone, 460 millimeters (18 inches) in diameter, with a hole in the center through which a metal rod 2.63 meters (8.6 feet) long was driven into the ground (Figure 11.4). A vertical micrometer was fastened to the stone. As the stone gradually sank by the burrowing action of the worms, the micrometer measured the distance by registering it against the top of the rod (Cattermole and Wolfe 1987). A substitute for the original worm stone can still be seen today at Down House. Charles Darwin began these experiments in 1877, and they were continued for 19 years, extending long after

FIGURE 11.4 The worm stone, designed by Horace Darwin, at Down House.

Charles's death. The results were published in the *Proceedings of the Royal Society* (H. Darwin 1901). Winter readings from January 1878 to March 1886 indicated that the stone sank 17.8 millimeters in eight years, a rate of 2.22 millimeters per year. This would be the equivalent of a little less than 1 inch every 10 years. In his earthworm book, Charles Darwin had reported that small stones on the surface sank at the rate of about 2.2 inches in 10 years (C. Darwin 1881, 142), although small pieces of gravel would sink much more rapidly than the large worm stone.

HORACE AND IDA

Horace and Emma Cecilia Farrer, known as "Ida," had fallen in love, although Raverat (1952) hinted that it was Uncle Lenny who was first attracted to Ida in all her "shining Victorian

FIGURE 11.5 *Left*: Forty-six-year-old Horace Darwin, circa 1897. *Right*: Ida Darwin. Cambridge University Library.

perfection." She was the daughter of Sir Thomas Henry Farrer, a rather pompous barrister and civil servant, and Frances ("Fanny") Erskine, his first wife. His second wife was Katherine Euphemia Wedgwood, the daughter of Hensleigh (Emma's brother) and Fanny Wedgwood (Healey 2001). Ida was Farrer's only daughter, and he wanted to make a good marriage match for her. He thought Horace was too weak and had no career. Farrer desired a more worldly man like himself for a son-in-law—a banker or barrister. This was embarrassing for Charles and Hensleigh. The young couple could not be dissuaded, however. Charles assured Farrer that Horace would inherit enough money to live comfortably (Loy and Loy 2010), giving Horace £5,000 worth of railway stock as proof (Desmond and Moore 1991). The wedding took place on 3 January 1880 at St. Mary's Church in London. The weather was chilly and so was the reception, as the families were not speaking to one another. The relationship between Farrer and Ida was strained for some time after her marriage (Loy and Loy 2010).

Horace and Ida settled in Cambridge, and Charles and Emma visited them as often as possible, given their age and

Charles's health. In his earthworm book, Charles cited Farrer's excavation of Roman ruins on Sir Thomas's property. Gwen Raverat (1952) mentioned that as a child she felt intimidated by Aunt Ida's perfection, compounded by the Darwins' roughness, but as Gwen grew older she came to "love her exceedingly" (Figure 11.5).

THE CAMBRIDGE SCIENTIFIC INSTRUMENT COMPANY

In 1878 Horace entered into a partnership with Albert G. Dew-Smith, a wealthy friend from student days, to make instruments for scientific research. George Darwin was also involved in the development of some of the instruments used for biological research. Horace designed a series of anthropometric instruments used by his cousin Francis Galton. One of the most important devices that he designed was the rocking microtome. It allowed biologists to quickly prepare microscopic specimens for examination. In keeping with the Darwin-Wedgwood interest in photography, early in 1880 Horace invented a light meter (actinometer) that was sensitive to infrared radiation (Cattermole and Wolfe 1987).

Horace became the chief shareholder in 1880, and on 1 January 1881 the Cambridge Scientific Instrument Company (CSIC), "Horace Darwin's shop," came into being (Cattermole and Wolfe 1987). He was in sole control by 1891, when Dew-Smith retired from the instrument side of the business (Glazebrook 2004).

The partners turned their attention to the reproduction of illustrations for scientific journals. They produced gorgeous plates for *Transactions of the Royal Society*. In 1883 the Indonesian island of Krakatoa erupted with a violence that humans had rarely witnessed. Dust clouds circled the earth and spectacular sunsets were seen for months around the globe. The Krakatoa Committee of the Royal Society published six plates illustrating the color of the English sky seen in November 1883 (Symons 1888). These plates are considered the finest examples of the chromolithography produced by the Cambridge Scientific Instrument Company (Cattermole and Wolfe 1987). Dew-Smith took over the lithographic printing business as the Cambridge Engraving Company. This lithographic business was later absorbed by Cambridge University Press. Dew-Smith died in 1903 (Cattermole and Wolfe 1987).

The Darwin family invested in the CSIC. Leonard Darwin became a director of the company, and the first share- and bondholders included Horace, Leonard, Francis, George, William, and Elizabeth Darwin (Cattermole and Wolfe 1987). Horace Darwin's shop produced instruments for his scientific friends at Cambridge as well as apparatuses for use in schools, colleges, and industry. Horace demonstrated a real talent for design and was responsible for a striking improvement in British scientific instrumentation (Glazebrook 2004). In Horace's obituary in *Nature*, R. T. Glazebrook (1928) wrote: "It was soon recognized that we had at Cambridge a firm of instrument makers the work of which would bear comparison with any in the world, while the head of the firm was a man with a genius for design and a knowledge of mechanics which enabled him to express his design in the simplest form consistent with the purpose for which the instrument was intended."

DARWIN GRANDCHILDREN, INCLUDING NORA BARLOW

Charles and Emma journeyed by special railway carriage to Cambridge to stay with Horace and Ida in the summer of 1881.

This greatly lifted Charles's spirits (Healey 2001). They received an early Christmas gift when their second grandchild arrived on 7 December 1881. Horace and Ida named him Erasmus, a revered Darwin name. Sadly, he was killed on 24 April 1915 during World War I at Ieper (Ypres) in western Belgium, the site of some of the fiercest fighting of the war.

Ida and Horace also provided two Darwin granddaughters, born after Charles's death. Ruth Frances Darwin arrived on 2 August 1883, married Rees Thomas, and lived for 90 years. Emma Nora Darwin was born on 22 December 1885. She married Sir James Allen Barlow and, under the name Nora Barlow, is well known as Charles Darwin's granddaughter. She transcribed Darwin's 18 H.M.S. *Beagle* diaries (Barlow 1933). This work is an important companion to *The Voyage of the Beagle.* Nora edited *Charles Darwin and the Voyage of the Beagle* (1945), *The Autobiography of Charles Darwin, 1809–1882, with Original Omissions Restored* (1958), and *Darwin and Henslow: The Growth of an Idea* (1967). She may be considered the originator of the "Darwin industry" through her careful study of Charles's manuscripts, the transcription of *The Voyage of the Beagle*, and the restoration of omissions made by Francis (at the behest of Emma Darwin) to Charles's *Autobiography* (Healey 2001). Nora took a course on "variation and heredity" from William Bateson at Cambridge in 1906. She was a friend of Ronald A. Fisher and did hybrid experiments with flowers, including *Aquilegia.* A cultivar of this columbine (*Aquilegia vulgaris* "Nora Barlow") is named after her (Richmond 2007; Browne 2010).

SOCIAL CONSCIENCE AND PUBLIC SERVICE

On the news of his father's death in 1882, Horace rushed to Down House to join the other family members for the funeral.

FIGURE 11.6 Horace Darwin as mayor of Cambridge, 1896–1897. Cambridge Scientific Instrument Company.

In addition to his strong family feelings, Horace was devoted to public service and was one of the founding members of the Cambridge Association for the Protection of Public Morals. In 1885, during its first year of operation, the association shut down eight brothels (Cattermole and Wolfe 1987). Horace became a justice of the peace and served on various civic and university boards. He raised £5,000 for an engineering laboratory at Cambridge University, and his political maneuvering was instrumental in the creation of this modern engineering school (R. S. W. 1928; Cattermole and Wolfe 1987).

In 1895 Horace was elected to a seat on the town council, and the following year he became mayor of Cambridge, holding that office from 1896 to 1897 (Figure 11.6). His connections with Cambridge University helped unite "town and gown" by improving the relationship between the city of Cambridge and the university. Horace was interested in seismological work and was appointed a member of the Seismological Committee of the British Association for the Advancement of Science. He and George designed a type of seismograph to record very small disturbances.

NEW THINGS

When a telephone exchange was installed in Cambridge around 1885, all the Darwins living in the area became subscribers, with Horace as No. 17, George as No. 10, and Francis as No. 18 (Cattermole and Wolfe 1987). Horace and George remained enthusiastic cyclists all their lives, despite numerous falls. Niece Gwen always considered Uncle Horace to be the sickliest of the Darwin brothers, with his ill heath rivaling that of her grandfather (Raverat 1952). Horace underwent successful surgery for appendicitis in 1893, when the procedure was very new and risky.

ANOTHER DARWIN F.R.S.

Horace was elected a Fellow of the Royal Society in 1903, in recognition of the significance of his work in the field of scientific instrumentation. Freeman (1984) noted that the Darwins have been Fellows of the Royal Society in a father-to-son sequence longer than that of the members of any other family. The streak ran from Dr. Erasmus Darwin, elected in 1761, to George's son, Sir Charles Galton Darwin (1887–1962), elected in 1922, a span of 201 years that encompassed five generations and seven fellows (Berra et al. 2010a). Charles was extremely proud of his sons, and in 1876 he wrote, "Oh Lord, what a set of sons I have, all doing wonders!" (Healey 2001).

WAR EFFORT LEADS TO SIR HORACE

World War I provided many opportunities for the CSIC. Its designers developed equipment to locate the position of enemy guns, based on measuring the time interval between the arrival of the sound of gunfire at different points on a measured base. Each unit consisted of six microphones spread out along a baseline of 9,000 yards (Cattermole and Wolfe 1987). Members of the firm were also exploring anti-submarine warfare at a secret workshop known as the "Skating Rink," where magnetic mines were developed. Horace and his company became involved in aeronautics and built wind tunnels for testing airframes. They also developed aircraft instruments, including a device to aid the pilot in maintaining a straight course in a fog. Horace helped create a height finder for determining the height and position of an object in the sky (Glazebrook 1928). He developed instruments, such as gunsights, for the new conditions of air warfare (R. S. W. 1928). He also designed and built a camera for photographing the stars. The CSIC emerged from World War I in a very prosperous condition. Horace was knighted (Knight Commander of the Most Excellent Order of the British Empire, or KBE) in 1918 as a result of his service on the Air Inventions Committee.

TRADEMARK

On 7 February 1919, Horace proposed a trademark of his own design for the company (Cattermole and Wolfe 1987). It consisted of a cam (a device that converts a rotary motion to a to-and-fro motion) enclosed in an electrical bridge (Figure 11.7). The elegant simplicity of this design combined mechanical and electrical precision with a brilliant play on words. This trademark became the symbol that associated the highest standards

FIGURE 11.7 Trademark of the Cambridge Scientific Instrument Company, designed by Horace Darwin, featuring a cam and an electrical bridge, thereby symbolizing Cambridge.

of technical excellence with the name Cambridge. It was also indicative of Horace's genius in simplifying complexity, the trait that made him a great designer.

The authors of a history of the Cambridge Scientific Instrument Company offered a fitting tribute to Horace in the first chapter of *Horace Darwin's Shop*: "Horace Darwin rarely charged realistic prices for the instruments which he designed and manufactured. His interest lay in finding the most elegant solutions to the engineering problems posed to him as an instrument designer" (Cattermole and Wolfe 1987). In 1924 the firm went public, and its name was changed to the Cambridge Instrument Company.

THE LAST YEARS

Horace developed an interest in the training of mentally handicapped children and established a children's home (Littleton House, in the village of Girton) with the help of his daughter Ruth (W. 1928; Cattermole and Wolfe 1987). In 1923 he endowed a student scholarship at Cambridge for the study of mental ailments, in remembrance of his son (*Times* [London] 1928; W. 1928).

The longtime home of Horace and Ida on Huntingdon Road in Cambridge was called the Orchard, because of its fruit trees. Horace bought this property from a larger parcel of land, the Grove, owned by his brothers William and George (Cattermole and Wolfe 1987). Raverat (1952) described Aunt Ida's house as "almost too full of detail; but beautiful, not grand—and every tiny corner of it loved and finished and exquisite." Horace died there on 22 September 1928, at the age of 77. He had never really retired from the "shop" but placed more and more responsibility for running the business on his managers and directors. His obituary in the *Times* [London] (24 September 1928) is subtitled "A Great Inventor." Burial was at St. Giles's cemetery in Cambridge, not far from his home. Ida lived until 1946.

GWEN RAVERAT'S IMPRESSIONS

Even when Gwen Raverat was a child, she recognized Uncle Horace's love of machinery. She noted that "his absorption in his dear machines always remained a barrier to me for they are not at all in my line; though I liked to watch the affection in his face and the tender movements of his beautiful sensitive hands as he touched them."

At the end of Raverat's (1952) reminiscences about the Darwin brothers (her uncles and father), she wrote: "When I read over what I had written about these five brothers, I felt that it might seem that I had made them too good, too nice, too single-hearted to be true. But it *was* true, for in a way that was what was wrong with them. I always used to feel that they needed protecting and cherishing, for they never seemed to me to have quite grown up.... They were quite unable to understand the minds of the poor, the wicked, or the religious.... But my grandfather was so tolerant of their separate individualities, so broad-minded, that there was no need for his sons to break away from him; and they lived all their lives under his shadow, with the background of the happiest possible home behind them.... At any rate, I know that I always felt older than they were. Not nearly so good, or so brave, or so kind, or so wise. Just older."

Gwen worked for a brief time in 1941 as a factory hand at the Chesterton Road branch of the Cambridge Instrument Company as part of the war effort, before being transferred to the Admiralty as a draughtswoman (Cattermole and Wolfe 1987). As late as 1958 the Darwin family still retained about 30 percent of the shares in the Cambridge Instrument Company (Cattermole and Wolfe 1987). On 14 June 1968, Horace Darwin's shop was sold to George Kent Ltd., with the support of the government's Industrial Reorganization Council.

CHARLES WARING DARWIN

(1856–1858)

On 13–16 April 1856, Charles Lyell visited Darwin at Down House. Lyell and Darwin discussed a paper written by a young British naturalist, Alfred Russel Wallace, that had appeared in the *Annals and Magazine of Natural History* the previous year. It seemed to present ideas that were similar to Darwin's on how races of animals may develop into species. The essence of Wallace's thesis was succinctly stated at the end: "Every species has come into existence coincident both in space and time with a pre-existing closely allied species." How this could happen, however, had not yet occurred to Wallace. Lyell cautioned Darwin to hurry up and publish his own ideas before someone beat him to it. Darwin appeared unmoved by this warning. He had read the paper and did not consider Wallace to be a threat. Darwin proceeded to explain his thoughts about natural selection to Lyell in depth, and Lyell continued to urge their publication (Browne 1995).

A TENTH CHILD

In spring 1856, Emma was stunned when she discovered that she was pregnant once again, although she thought she may have had a couple of miscarriages in the intervening years

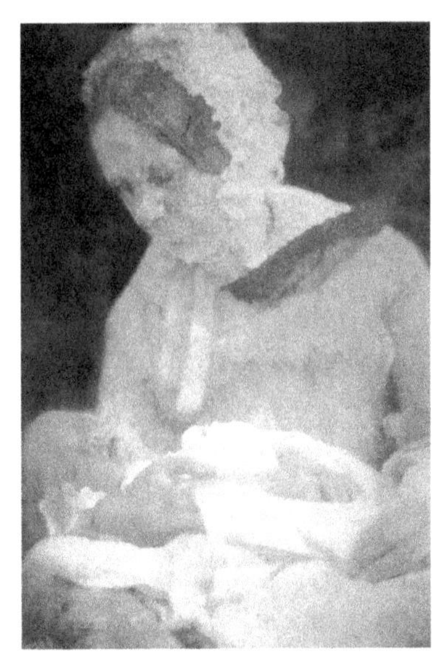

FIGURE 12.1 Emma Darwin holding Charles Waring Darwin, photographed by William Darwin, in 1857. English Heritage Photographic Library, London.

(Healey 2001). She would turn 48 years old on 2 May, and she had already given birth nine times. Her previous child, Horace, was born five years earlier. Emma was as wretchedly uncomfortable as ever (Burkhardt et al. 1985–, 6: 87, 151–152, 191). In 1852 Charles had written to his cousin and Cambridge University classmate, William Darwin Fox, to congratulate him on the birth of Fox's tenth child. Charles then added, "but please to observe when I have a 10th, send only condolences to me" (Burkhardt et al. 1985–, 5: 83).

The Darwin's tenth and last child was born on 6 December 1856. Emma's sister Elizabeth was present to help, as she had been many times before. Charles administered chloroform, but not nearly as much as he had in the past, and only when Emma cried out for it (Burkhardt et al. 1985–, 6: 438). The baby was christened Charles Waring on 21 May 1857, at the parish church in Downe. It was apparent that the baby was not normal. The probability of a Down syndrome baby rises with increasing maternal age (Huether et al. 1998). The risk is about 1 in 14 for a 48-year-old mother (Beers and Berkow 1999)—Emma's age then—compared to 1 in 1,667 for a 20-year-old mother. R. Keynes (2001) suggested a Down syndrome diagnosis for this last Darwin child. A photograph of Emma holding baby Charles Waring, taken by William, displays typical Down syndrome features (Figure 12.1). The skull shows frontal bossing; the eyes are wide set, with a hint of slanting; the nasal bridge is depressed; and the lips are thin. The hands appear to have short, wide digits, and there is a suggestion of clinodactyly in the fifth finger. The baby appears hypotonic (Berra et al. 2010a).

In today's parlance, Charles Waring was "developmentally delayed." He showed no sign of walking or talking, but was "'remarkably sweet' and affectionate, with a 'wicked little smile,' he was totally passive, and 'made strange grimaces & shivered, when excited.' He had a 'passion for Parslow' and a special claim on his mother" (Desmond and Moore 1991).

WALLACE'S LETTER

In early March 1858, while recovering from an attack of malaria on Ternate Island in the Moluccas (= Spice Islands, between Sulawesi and New Guinea, now a province of Indonesia), Wallace decided to write a letter to Darwin that more fully outlined his views on species. Wallace chose to do this because, in previous correspondence (1 May 1858), Darwin had written to Wallace: "By your letter [not found] & even still more by your paper in Annals [Wallace 1855] . . . I can plainly see that we have thought much alike & to a certain extent have come to similar conclusions. . . . I agree to the truth of almost every word of your paper" (Burkhardt et al. 1985–, 6: 387–388).

This pleased Wallace greatly, and he wrote to his fellow Amazon explorer, Henry Walter Bates, on 4 January 1858: "I have been much gratified by a letter from Darwin, in which he says that he agrees with 'almost every word' of my paper. He is now preparing his great work on 'Species and Varieties' for which he has been collecting material for twenty years. He may save me the trouble of writing more on my hypothesis, by proving that there is no difference in nature between the origin of species and varieties; or he may give me trouble by arriving at another conclusion; but at all events, his facts will be given for me to work upon" (Wallace 1905, 1: 358). The Darwin-Wallace relationship, as reflected in their correspondence, has been reviewed by Berra (2013).

On 18 June 1858, Charles finally received the Ternate letter from Wallace, which contained a manuscript entitled "On the Tendency of Varieties to Depart Indefinitely from the Original Type." Like Darwin had done 20 years previously, Wallace had read Thomas Malthus's *Essay on the Principle of Population.* This led Wallace to the idea of natural selection, just as it had done for Darwin. Darwin had 20 years of data to support this concept; Wallace intuitively came to the same conclusion, but without the data to back it up. Wallace asked that Darwin read the manuscript and pass it along to Charles Lyell if Darwin considered it significant (Wallace 1905, 1: 362–363).

Darwin wrote to Lyell the same day, enclosing Wallace's manuscript (as Wallace had requested), saying: "Your words have come true with a vengeance—that I should be forestalled." Charles further stated that "if Wallace had my MS. Sketch written out in 1842, he could not have made a better short abstract" (Burkhardt et al. 1985–, 7: 107).

VESTIGES

Charles Lyell had been trying to get Darwin to publish his ideas on evolution for several years, before someone else arrived at the same conclusion, but Darwin was reluctant to rush into print. He had many excuses for procrastinating. Charles knew he would be attacked by religious authorities and believers for proposing a godless origin for the earth's various species. He didn't want to upset devout Emma. He wanted to make sure he could answer every possible objection by completing his "big book," *Natural Selection*. He had seen the upset caused by *Vestiges of the Natural History of Creation*, published anonymously by Robert Chambers in 1844 (Secord 2000). Chambers posited a view of evolution as being driven by innate forces that caused species to strive for change. The book was wildly popular—but full of scientific errors. Because it was published anonymously, there was much speculation about its authorship. Adam Sedgwick, Darwin's Cambridge geology professor, initially thought that because it was so awful, it must have been written by a woman (imagine saying that today in a book review), but Sedgwick later changed his opinion (Sedgwick 1845). Public acknowledgment of Chambers's authorship was not made until 1884, after Darwin's death, but Charles suspected that Chambers was the author (Freeman 1978).

DEATH OF CHARLES WARING DARWIN

Henrietta was sick with fever (diphtheria), three children in the village of Downe had died recently from scarlet fever, and baby Charles was very ill with fever. On 28 June 1858, Charles Waring succumbed. He was buried on 30 June, next to infant Mary in the parish churchyard, with Charles and Emma present. As fears of a local epidemic spread, the next day Darwin evacuated most of his family to Sarah Elizabeth Wedgwood's (Emma's sister) house in Sussex. Charles and Emma remained at Down to take care of the ailing Henrietta (Healey 2001). Years later, in her role as editor of her mother's letters, Henrietta wrote: "The poor little baby was born without its full share of intelligence. Both my father and mother were infinitely tender towards him, but, when he died in the summer of 1858, after their first sorrow, they could only feel thankful. He had never learnt to walk or talk" (H. Litchfield 1915, 2: 162).

TO PUBLISH OR NOT TO PUBLISH

Darwin agonized over the morality of publishing his 20 years' worth of data, now that he had received Wallace's essay also positing the principle of natural selection. Charles did not want to be accused of acting unethically and claiming priority over an amateur naturalist. Shermer (2002) has explored this minefield in great historical and philosophical detail and judged that Darwin acted as the perfect gentleman that he was. This conclusion is clearly justified by the letter Darwin wrote to Charles Lyell one week after receiving Wallace's letter. In this piece of correspondence, dated 25 June 1852, Darwin expressed to Lyell that he felt constrained by Wallace's letter:

> My Dear Lyell
>
> I am very sorry to trouble you, busy as you are, in so merely personal an affair. But if you will give me your deliberate opinion, you will do me as great a service, as ever man did, for I have entire confidence in your judgment & honour....
>
> There is nothing in Wallace's sketch which is not written out much fuller in my sketch copied in 1844, & read by Hooker some dozen years ago. About a year ago I sent a short sketch of which I have copy of my views (owing to correspondence on several

points) to Asa Gray, so that I could most truly say & prove that I take nothing from Wallace. I sh[d] be *extremely* glad **now** to publish a sketch of my general views in about a dozen pages or so. But I cannot persuade myself that I can do so honourably. Wallace says nothing about publication, & I enclose his letter.—But as I had not intended to publish any sketch, can I do so honourably because Wallace has sent me an outline of his doctrine?—I would far rather burn my whole book, than that he or any other man sh[d] think I had behaved in such a paltry spirit. Do you not think that his having sent me this sketch ties my hands?... By the way would you object to send this & your answer to Hooker to be forwarded to me, for then I shall have the opinion of my two best & kindest friends.— This letter is miserably written and I write it now, that I may for time banish whole subject. And I am worn out with musing.

AN HONORABLE SOLUTION

Lyell devised an ingenious solution to prevent his friend from losing priority over the topic he had researched for the past 20 years: Darwin and Wallace would announce their ideas together. Darwin had the concept fully formed by 1844. Joseph Dalton Hooker had looked at Darwin's 1844 essay, and Asa Gray from Harvard had received extracts from it. This demonstrated that Darwin borrowed nothing from Wallace. Darwin sent his Asa Gray letter to Hooker and left the matter in the hands of his friends Lyell and Hooker. He was too distraught over his baby's recent death to deal with Wallace's letter himself. Darwin wrote to Hooker on 29 June (Burkhardt et al. 1985–, 7: 121–122):

My dear Hooker

I have just read your letter, & see you want papers at once. I am quite prostrated & can do nothing but I send Wallace & my

> abstract of abstract of letter to Asa Gray, which gives most imperfectly only the means of change & does not touch on reasons for believing species do change. I daresay all is too late. I hardly care about it—
>
> But you are too generous to sacrifice so much time & kindness.—It is most generous, most kind. I send sketch of 1844 solely that you may see by your own handwriting that you did read it.—
>
> I really cannot bear to look at it.—Do not waste much time. It is miserable in me to care at all about priority.—
>
> The table of contents will show what it is. I would make a similar, but shorter & more accurate sketch for Linnean Journal.—I will do anything
>
> God bless you my dear kind friend. I can write no more. I send this by servant to Kew.
>
> Yours
>
> C. Darwin

The Darwin and Wallace papers were combined alphabetically and inserted into a hastily organized meeting of the Linnean Society on 1 July 1858. Extracts of Darwin's 1844 manuscript, portions of his 1857 letter to Gray, and Wallace's manuscript were read by the society's secretary, John Joseph Bennett (Desmond and Moore 1991; Browne 2002). This material was not presented as one paper, with Darwin and Wallace as coauthors. Rather, the items were treated as two separate contributions, each with its own author, and were presented as a unit. About 30 people were present at this meeting, but no great excitement was generated by the papers (Moody 1971). The audience didn't "get it." It would take a book to explain this momentous concept. Grief-stricken and ill himself, Darwin did not attend the meeting, but instead remained at Down to deal with his baby son's funeral.

When word of the Linnean Society meeting and the reading of the joint papers finally reached Wallace, who was in the

Malay jungle, by letters to him from Darwin and Hooker, he was very pleased. In the second edition of his autobiography (Wallace 1908, 193), Wallace wrote that "I not only approved, but felt that they had given me more honour and credit than I deserved, by putting my sudden intuition... on the same level with the prolonged labours of Darwin, who had reached the same point twenty years before me." The joint papers were published in the *Journal of the Proceedings of the Linnean Society (Zoology)* in August 1858 (C. Darwin and Wallace 1858).

Van Wyhe and Rookmaaker (2012) presented a convincing explanation for the timing of Darwin's receipt of Wallace's letter—on 18 June 1858—by reviewing shipping and mail schedules, although Davis (2012) disputed their calculations. Porter (2012) further explored why Wallace contacted Darwin in the first place. Berra (2013) examined the Darwin-Wallace correspondence and, using Wallace's own words, demonstrated Wallace's unequivocal acceptance of Darwin's priority with grace and a complete lack of acrimony.

MEMORIAL TO CHARLES WARING DARWIN

On 2 July 1858, Charles Darwin wrote a touching memorial to his infant son, who had lived less than 19 months (Burkhardt et al. 1985–, 7: 521):

> "Our poor Baby was born Dec[r] 6th 1856 & died on June 28th 1858, & was therefore above 18 months old. He was small for his age & backward in walking & talking, but intelligent & observant. When crawling naked on the floor he looked very elegant. He had never been ill, & cried less than any of our babies. He was of a remarkably sweet, placid & joyful disposition; but had not high spirits, & did not laugh much. He often made strange grimaces &

shivered, when excited; but did so also, for a joke & his little eyes used to glisten, after pouting out or stretching widely his little lips. He used sometimes to move his mouth as if talking loudly, but making no noise, & this he did when very happy. He was particularly fond of standing on one of my hands, & being tossed in [the] air: & then he always smiled, & made a little pleased noise. I had just taught him to kiss me with open mouth, when I told him. He would lie for a long time placidly on my lap looking with a steady & pleased expression at my face; sometimes trying to poke his poor little fingers into my mouth, or making nice little bubbling noises as I moved his chin. I had taught him not to scratch, but when I said "Giddlums never scratches now" he could not always resist a little grab, & then he would look at me with a wicked little smile. He would play for any length of time on the sofa, letting himself fall suddenly, & looking over his shoulder to see that I was ready. He was very affectionate, & had a passion for Parslow; & it was very pretty to see his extreme eagerness, with outstretched arms, to get to him. Our poor little darling's short life has been placid innocent & joyful. I think & trust he did not suffer so much at last, as he appeared to do; but the last 36 hours were miserable beyond expression. In the sleep of Death he resumed his placid looks."

ON THE ORIGIN OF SPECIES

In July Charles took Emma and Henrietta to join the family, who were "in exile" at Emma's sister's house in Hartfield. The serenity and sea air revived all concerned. They returned to Down House on 13 August, and Charles began to work on his theory (Healey 2001). The close call with Wallace was enough to stimulate Darwin to abandon his "big book" and complete what he called "an abstract" of his idea about species. The rest of the world calls it *On the Origin of Species*. It was published on 24 November 1859. Modern biology had begun.

The Origin was an immediate success. All 1,250 copies of the first edition were spoken for before the publication date, and 3,000 copies of the second edition were rushed into print and quickly sold (Freeman 1977).

HUXLEY-WILBERFORCE DEBATE

On 30 June 1860, *The Origin* was the topic of discussion at the British Association for the Advancement of Science meeting in Oxford, with John Henslow in the chair as president. The bishop of Oxford, Samuel Wilberforce, derided the book and addressed Thomas Henry Huxley—who was representing his absent friend Charles Darwin—and asked if it was through Huxley's grandfather or his grandmother that he descended from an ape. Huxley replied with words to the effect that he would rather have an ape for a grandfather than a bishop who ridiculed science. The meeting dissolved into chaos and this temporarily silenced the bishop. It also earned Huxley the nickname of "Darwin's Bulldog." Game on—the battle between evolution and religion had begun! Huxley's subsequent success in public forums can be traced back to this moment (Meacham 1970, 216–217).

The above description is a truncated caricature of the Oxford meeting. The exact words of the debate were not recorded at the time. In her biography of Darwin, Browne (2002) gave a full account, rich in details. Joseph Dalton Hooker wrote a description of the Oxford meeting just two days later (on 2 July), for Darwin's benefit (Burkhardt et al. 1985–, 8: 270–271). He described the proceedings and how both Huxley and himself had defended their friend's theory. Browne (1978) examined Hooker's account in an analysis of the prolific and historic Darwin-Hooker correspondence. There are also several eyewitness accounts that differ in the wording of the exchange between Wilberforce and Huxley. Lucas (1979) attempted to rehabilitate Wilberforce, and Gould (1986) downplayed Huxley's triumph.

HUXLEY'S AND WILBERFORCE'S WORDS

Two months after the debate, Huxley wrote a letter to his Welsh surgeon-naturalist friend, Frederic Daniel Dyster, in which Huxley recounted his words at that extraordinary meeting: "If then, said I, the question is put to me would I rather have a miserable ape for a grandfather or a man highly endowed by nature and possessed of great means and influence and yet who employs those faculties for the mere purpose of introducing ridicule into a grave scientific discussion—I unhesitantly affirm my preference for the ape" (Burkhardt et al. 1985–, 8: 271–272).

Years later, the meeting was described in Darwin's *Life and Letters* (F. Darwin 1887, 2: 320–323) and in Huxley's *Life and Letters* (L. Huxley 1900, 1: 179–189). Wilberforce reportedly said, "If any one were to be willing to trace his descent through an ape as his *grandfather*, would he be willing to trace his descent similarly on the side of his *grandmother*?" (L. Huxley 1900, 183). Huxley allegedly turned to Sir Benjamin Brodie, the president of the Royal Society who was seated next to Huxley, and whispered, "The Lord hath delivered him into mine hands" (L. Huxley 1900, 184). Huxley then stated: "I asserted—and I repeat—that a man has no reason to be ashamed of having an ape for his grandfather. If there were an ancestor whom I should feel shame in recalling it would rather be a *man*—a man of restless and versatile intellect—who, not content with an equivocal success in his own sphere of activity, plunges into scientific questions with which he has no real acquaintance, only to obscure them by an aimless rhetoric, and distract the attention of his hearers from the real point at issue by eloquent digression and skilled appeals to religious prejudice" (L. Huxley 1900, 185).

More recently, Desmond (1997, 276–281) and Jensen (1988) recounted various versions of this epic confrontation. By comparing all the sources cited above, which more or less tell the same story, one can capture the essence and excitement of the moment, even though the precise wording is elusive. The Oxford debate scenario has become such a cultural icon of the conflict between science and religion that it is even included in *Bartlett's Familiar Quotations* (Bartlett 1968, 725b; Anderson 2002).

As usual, the acerbic Huxley had the last word. When, in 1873, the bishop fell off his horse and died, Huxley wrote to a friend: "Poor dear Sammy! His end has been all too tragic for his life. For once, reality & his brains come into contact & the result was fatal" (Desmond 1997). *The Origin* went through six editions. and "*On*" was dropped from the title after the first edition (Costa 2009).

LYELL'S AND HUXLEY'S BOOKS

In February 1863, Darwin's close friend and confidant, geologist Sir Charles Lyell, who had so helpfully brokered the compromise arrangement with Wallace's paper, published *The Geological Evidences of the Antiquity of Man with Remarks on Theories of the Origin of Species by Variation*. The publisher was Darwin's very own: John Murray of London. Although Lyell accepted evolution, he hedged about human origins and allowed theistic input. In fact, he required "a direct interposition of the Deity." This was a great disappointment to Darwin. After all, it was Lyell's geology that had started Darwin thinking about evolution during the voyage of the H.M.S. *Beagle*. In addition, Lyell's geology demonstrated the immense age of the earth, which provided enough time for evolution to occur. But Lyell just couldn't bring himself to go all the way with a naturalistic origin of humans.

Huxley, as expected, was not so bashful. His book, *Evidence as to Man's Place in Nature*, was published shortly after Lyell's *Antiquity of Man*. As a morphologist, Huxley compared the anatomy of humans and other apes (gibbons, orangutans, gorillas, and chimpanzees), monkeys, and lemurs. He concluded that there was no reason to doubt that man could have originated, by means of natural selection, from an apelike ancestor. This, of course, was in the days before the extensive fossil record of human ancestors was known to confirm Huxley's view. Charles was heartened by Huxley's little "monkey book," as Darwin called it.

EPILOGUE

By any standard, Charles and Emma Darwin created a loving, nurturing environment for their 10 offspring. The children were devoted to their father and mother, intensely loyal to the family and to each other, and protective of their father's reputation.

Three of the children died young: Annie (age 10), Mary Eleanor (three weeks), and developmentally delayed Charles Waring (1.5 years). Annie's death was devastating to her parents, and Charles reflected on this loss for the rest of his life. He wrote a loving tribute to Annie, and a fond memorial to his toddler son. Mary's too-short life precluded any such encomium for her.

The Darwin household was an incredibly stimulating place in which to grow up. The children were exposed to nature walks, led by their father, that were replete with biological observations; vacations at the sea shore; and visits to London that were enhanced with the excitement of the zoo, the botanical garden, museums, and exhibitions. They assisted their father in breeding pigeons and in experimenting with plants in the greenhouse and garden. Numerous dogs, kittens, horses, cows, birds, and other domestic animals were available for play and learning.

Charles provided an excellent role model in persistence, by means of his eight-year barnacle studies in his home office. Yet the children were not directed or forced into science. They simply absorbed it. It was a routine part of their life, their normal reality. Of course, it didn't hurt that the family was very wealthy. Expensive hobbies such as photography and mechanical tinkering could be encouraged, and several of the boys became highly adept, later using their skills in their careers.

In the custom of the time, the daughters were educated at home, and this task was joyfully undertaken by Emma. She read to all of her children, and this tradition was continued by Henrietta and Elizabeth. William, the firstborn son, was sent to prep school for a classical education. Charles came to suspect that this was not the best training for the development of a child's mind. His succeeding sons were all sent to a school near home that instead prepared them for any path they might choose.

Charles was proud of the accomplishments of his surviving children. Each was a unique personality. They were given the freedom to be whatever they could become. They were not expected to follow in their famous father's footsteps, and they were not envious of each other. The Darwin upbringing infused these offspring with a sense of wonder and curiosity, as well as a scientific methodology that encouraged observation, experimentation, and analysis. This, no doubt, influenced their career paths.

All of the sons except Leonard were Cambridge men like their father. William spent his adult life as a banker, and successfully managed the family's substantial wealth. Henrietta fulfilled the role of Charles's secretary and editor, and he relied on her to polish his work. George was probably the most distinguished of the Darwin children, with immense achievements in geophysics and astronomy. Francis was his father's secretary, lab assistant, coauthor, and biographer. In his spare time, Francis practically invented the fields of plant physiology and the study of plant hormones. Leonard, educated at the Royal Military Academy, was the most broadly based of the Darwin siblings. He exerted a prominent influence in the army, as well as on economics, politics, and eugenics. Horace and his Cambridge Scientific Instrument Company were vitally important to England's scientific reputation and, eventually, to the war effort during World War I. George, Francis, and Horace were knighted and elected Fellows of the Royal Society. While Charles was aware of some of these honors, how proud he would have been to have known about them all.

As adults, most of the Darwin children gravitated toward one another, eventually coming to live in the same general area of Cambridge. Charles put Emma on a pedestal, and the children kept her there. After their father's death, they all spent time with their mother, who eventually settled close to her Cambridge-based children. Elizabeth, who lived with her mother until Emma's death, then remained in Cambridge, near her brothers. Theirs was a close-knit family that genuinely loved one another. This is a reflection on Charles and Emma's devotion to each other and to the family, which, in turn, promoted the success and general likability of the Darwins.

Charles and Emma raised an extraordinary family, whose individual members accomplished far more than one would have the right to expect. I hope this little book makes that fact more widely known.

APPENDIX 1: TIMELINE

The timeline shows the chronological relationship of Darwin's biological children and his literary offspring.

TIMELINE PHOTOS

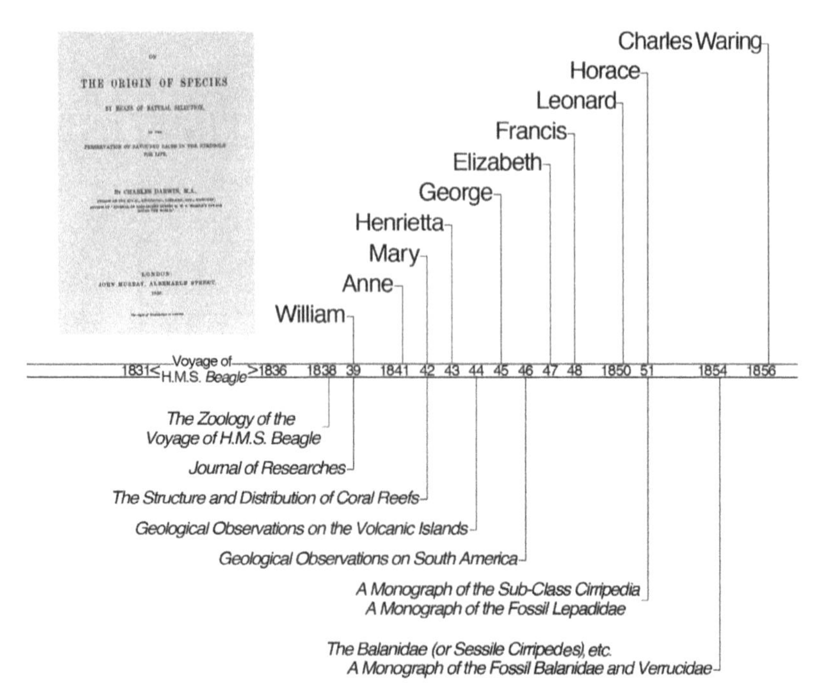

Top left: The title page from *On the Origin of Species*, published by John Murray on 24 November 1859.
Bottom left: The Darwin children on holiday at Caerdeon, North Wales, in 1869. *From left to right*: Henrietta, Francis, Leonard, Horace, and Elizabeth. Cambridge University Library.*Top right*: Four of the five Darwin sons with their Uncle Ras, in

1859 1862 1865 1868 1870 71 72 1875 76 77 1879 80 81

On the Origin of Species
Fertilisation of Orchids
On the Movements and Habits of Climbing Plants
The Variation of Animals and Plants under Domestication
The Descent of Man and Selection in Relation to Sex
The Expression of Emotions in Man and Animals
Insectivorous Plants
The Effects of Cross- and Self-Fertilisation
The Different Forms of Flowers
Erasmus Darwin
The Power of Movement in Plants
The Formation of Vegetable Mould, through the Action of Worms

the early 1870s. *From left to right*: Horace, Leonard, Erasmus Alvey Darwin (Uncle Ras), Francis, and William. Note William's striking resemblance to the young Charles Darwin. This identification agrees with Browne (2002) and Loy and Loy (2010). (Compare Leonard's suit in both group photos on this page.) Burkhardt et al. (1985–, 18: frontispiece) instead identified the sitting figure on the left as Leonard, and the person standing next to him as George. Cambridge University Library.

APPENDIX 2: CAST OF CHARACTERS

What follows is based on Benét (1965), Bowlby (1990), Burkhardt et al. (1985–), Freeman (1978), Ghiselin (2009), *The Encyclopaedia Britannica* 11th and 15th ed. and *The Oxford Dictionary of National Biography*.

Allen, Frances "Fanny." 1781–6 May 1875. Emma Darwin's (ED) unmarried aunt.

Allfrey, Charles Henry. 1838/1839–1912. Physician. Attended Charles Darwin (CD) in his terminal illness. Personal friend, invited to funeral of CD.

Balfour, Sir Arthur James. 25 Jul. 1848–19 Mar. 1930. Statesman. Fellow of Royal Society (FRS). Prime Minister, 1902–1905. Cambridge friend of CD's sons. Personal friend, invited to funeral of CD.

Barlow, Lady Emma Nora née Darwin. *See* Emma Nora Darwin.

Barlow, Sir James Alan Noel. 1881–1966. Civil servant. Husband of Emma Nora Darwin, known as Nora Barlow.

Bates, Henry Walter. 8 Feb. 1825–16 Feb. 1892. Naturalist, entomologist. Explored Amazon with A. R. Wallace, 1848–1850, and alone until 1859. Author of *Naturalist on the River Amazons*

(1863). Developed concept of protective coloration, now known as Batesian mimicry. Assistant secretary, Royal Geographical Society, 1864–1892. FRS, 1881.

Bennett, John Joseph. 8 Jan. 1801–29 Feb. 1876. Botanist. Secretary, Linnean Society, 1838–1852.

Brodie, Sir Benjamin Collins. 19 Jun. 1783–21 Oct. 1862. Surgeon. FRS, 1810. Professor of comparative anatomy and physiology, Royal College of Surgeons, 1816. Baronet, 1834. Consulted by Queen Victoria and ED. President, Royal Society, 1858–1861. Seated next to T. H. Huxley at Oxford debate with Bishop Samuel Wilberforce.

Brodie, Jessie. ?–1873. Scottish nurse for Darwin children at 12 Upper Grower Street and Down House, 1842–1851. Left after Annie Darwin's death.

Browning, Elizabeth Barrett. 6 Mar. 1806–29 Jun. 1861. Poet. Wife of Robert Browning. Author of *Sonnets from the Portuguese* (1850).

Browning, Robert. 7 May 1812–12 Dec. 1889. Poet. Husband of Elizabeth Barrett Browning. Author of *The Ring and the Book* (1868–1869).

Cameron, Julia Margaret. 11 Jun. 1815–26 Jan. 1879. Photographer. Photographed CD on Isle of Wight, 1868.

Carlyle, Jane Baillie née Welsh. 1801–1866. Wife of Thomas Carlyle.

Carlyle, Thomas. 4 Dec. 1795–5 Feb. 1881. Scottish-born writer and historian. Known for his explosive attacks on sham and hypocrisy. Author of *The French Revolution* (3 vols., 1837). Acquaintance of CD and Erasmus Alvey Darwin.

Chambers, Robert. 10 July 1802–17 Mar. 1871. Scottish publisher and writer. Anonymous author of *Vestiges of the Natural History of Creation* (1844). Its negative reception may have contributed to CD's reluctance to publish his ideas on evolution.

Chester, Joseph Lemuel. 1821–1882. American genealogist. Hired by George Darwin to compile Darwin family tree.

Clark, Sir Andrew. 28 Oct. 1826–6 Nov. 1893. Fashionable London physician. FRS, 1885. Attended CD in London and at Down House. Personal friend, invited to funeral of CD.

Cornford, Frances née Darwin. *See* Frances Crofts Darwin.

Cornford, Francis Macdonald. 1874–1943. Professor of ancient philosophy, Cambridge University. Husband of Frances Crofts Darwin. Father of poet Francis Cornford.

Cotton, Elizabeth Reid (aka Lady Hope). Writer of evangelical and temperance tracts. Created mythical story of CD's deathbed conversion. Widow of Admiral James Hope.

Covington, Syms. 1816–17 Feb. 1861. CD's personal servant on H.M.S. *Beagle* and afterward, 1833–1839. Settled in New South Wales. Sent specimens to CD from Australia.

Crofts, Ellen Wordsworth. 1856–1903. Second wife of Sir Francis Darwin (1883). Lecturer at Newnham College, Cambridge. Mother of Frances Crofts Darwin.

Darwin, Amy Richenda née Ruck. *See* Amy Richenda Ruck.

Darwin, Anne Elizabeth "Annie." 2 Mar. 1841–23 Apr. 1851. Second child, first daughter of CD and ED. Died at age 10.

Darwin, Bernard Richard Meirion. 7 Sept. 1876–18 Oct. 1961. Sportswriter for the *Times* [London] (1908–1953), golfer. Firstborn grandson of CD. Only child of Francis and Amy Ruck Darwin. After Amy died in childbirth, raised by ED at Down House until Francis remarried in 1882.

Darwin, Caroline Sarah. 14 Sept. 1800–5 Jan. 1888. CD's sister. Second child of Robert Waring Darwin. Wife of Josiah Wedgwood III (1837).

Darwin, Sir Charles Galton. 9 Dec. 1887–31 Dec. 1962. Physicist. CD's grandson. Second child of Sir George Howard Darwin. FRS, 1922. Knighted, 1942. Professor of natural philosophy, Edinburgh University, 1923–1936.

Darwin, Charles Robert. 12 Feb. 1809–19 Apr. 1882. Naturalist. FRS, 1839. Copley Medal, 1864. Codiscoverer of evolution by natural selection. Author of *On the Origin of Species* (1859), which marks the beginning of modern biology.

Darwin, Charles Waring. 5 Dec. 1856–28 June 1858. Tenth and last child of CD and ED. Most likely a Down syndrome baby.

Darwin, Charlotte Mildred née Massingberd. *See* Charlotte Mildred Massingberd.

Darwin, Elizabeth "Bessy." 8 Jul. 1847–8 June 1926. Sixth child, fourth daughter of CD and ED. Never married.

Darwin, Elizabeth Frances née Fraser. *See* Elizabeth Frances Fraser.

Darwin, Ellen Wordsworth née Crofts. *See* Ellen Wordsworth Crofts.

Darwin, Emily Catherine "Catty." 10 May 1810–2 Feb. 1866. CD's sister. Sixth child of Robert Waring Darwin. Second wife of Charles Langton.

Darwin, Emma née Wedgwood. See Emma Wedgwood.

Darwin, Emma Cecilia "Ida" née Farrer. See Emma Cecilia Farrer.

Darwin, Emma Nora. 22 Dec. 1885–29 May 1989. Author, editor. CD's granddaughter. Third child of Horace Darwin. Wife of Sir James Allen Barlow. Editor of *Charles Darwin and the Voyage of the* Beagle (1945) and *Autobiography of Charles Darwin* (1958), author of *Darwin and Henslow* (1967). Known as Lady Nora Barlow.

Darwin, Erasmus I. 12 Dec. 1731–17 Apr. 1802. Physician, naturalist, poet, philosopher. Grandfather of CD. FRS, 1761. Proposed an evolutionary concept later enlarged by Jean Baptiste de Lamarck. Author of *Botanic Garden* (1791), *Zoonomia* (1794), *Phytologia* (1800), and *Temple of Nature* (1803). Lengthy biographical sketch of his grandfather written by CD in 1879.

Darwin, Erasmus III. 7 Dec. 1881–24 Apr. 1915. CD's grandson, the second of two grandsons born during his lifetime. Son of Sir Horace Darwin. Unmarried. Killed in WWI.

Darwin, Erasmus Alvey "Ras." 29 Dec. 1804–26 Aug. 1881. London socialite. Older brother of CD. Second child of Robert Waring Darwin. Trained as physician, but never practiced. Close friend of Hensleigh Wedgwood, Thomas Carlyle, and Harriet Martineau.

Darwin, Florence Henrietta née Fisher. *See* Florence Henrietta Fisher.

Darwin, Frances Crofts. 30 Mar. 1886–1960. CD's granddaughter. Only child of Sir Frances and Ellen Wordsworth Darwin. Wife of Francis Macdonald Cornford. Mother of poet Francis Cornford.

Darwin, Sir Francis. 16 Aug. 1848–19 Sept. 1925. Botanist. Seventh child, third son of CD and ED. FRS, 1882. Knighted, 1913. Qualified as physician, but never practiced. Served as CD's assistant and secretary, 1874–1882. Junior author of *The Power of Movement in Plants* (1880). Reader in botany at Cambridge University, 1888–1904. Edited CD's *Life and Letters* (1887). Outlived three wives: Amy Ruck, married 1874–1876 (one son, Bernard); Ellen Crofts, married 1883–1903 (one daughter, Frances); and Florence Fisher, married 1913–1920. Coauthor of *Practical Physiology of Plants* (1894), author of *Elements of Botany* (1895).

Darwin, Sir George Howard. 9 Jul. 1845–7 Dec. 1912. Mathematician. Fifth child, second son of CD and ED. Plumian Professor of Astronomy, Cambridge, 1883–1912. FRS, 1879. Knighted, 1905. Author of *The Tides* (1898), and world's authority on tides. Author of many publications on earth, moon, sun system of tides, composition of earth, origin of moon. Husband of Maud du Puy of Philadelphia, PA. Father of Gwendolen Mary Darwin, Charles Galton Darwin, Margaret Elizabeth Darwin, and William Robert Darwin. Only remaining male line of CD's family comes through him.

Darwin, Gwendolen Mary. 26 Aug. 1885–1957. Artist. CD's granddaughter. Daughter of Sir George Howard Darwin. Wife of Jacques Raverat. As Gwen Raverat, author of *Period Piece* (1952), describing the lives of her grandparents, aunts, and uncles.

Darwin, Henrietta Emma. 25 Sept. 1843–17 Dec. 1927. Fourth child, third daughter of CD and ED. Sickly childhood. Assisted CD with editing and proofreading of many books, including *The Descent of Man* (2 vols., 1870–1871) and *Erasmus Darwin* (1879). Wife of Richard Buckley Litchfield (1871). No children. CD's only married daughter. Edited *Emma Darwin: A Century of Family Letters* (2 vols., 1904 and 1915) as Mrs. Litchfield.

Darwin, Sir Horace. 13 May 1851–22 Sept. 1928. Civil engineer. Ninth child, fifth and youngest surviving son of CD and ED. Founder and director of Cambridge Scientific Instrument Company, 1881. Mayor of Cambridge, 1896–1897. FRS, 1903. Knighted, 1918. Husband of Emma "Ida" Cecilia Farrer (1880). Father of Erasmus Darwin, Ruth Francis Darwin, and Emma Nora Darwin (CD's biographer, as Nora Barlow).

Darwin, Leonard. 15 Jan. 1850–26 Mar. 1943. Military engineer. Eighth child, fourth son of CD and ED. Army officer, economist, politician, eugenicist. Royal Engineers, 1871–1890, retired as major. Instructor in military engineering, chemistry, photography, 1877–1882. Intelligence Department, War Office, 1885–1890. Liberal-Unionist member of parliament (MP) for Litchfield, 1892–1895. President, Royal Geographical Society, 1908–1911. President, Eugenics Society, 1911–1928. Author of *Bimetallism* (1897) and *The Need for Eugenic Reform* (1926). Husband of Elizabeth "Bee" Frances Fraser (1882–1898). Second marriage to Charlotte Mildred Massingberd (1900–1940). No children from either marriage.

Darwin, Margaret Elizabeth. 1890–1974. CD's granddaughter. Third child of Sir George Howard Darwin. Wife of

Sir Geoffrey Keynes (1917). As M. E. Keynes (1943), wrote biographical sketch of her uncle, Leonard Darwin.

Darwin, Marianne. 7 Apr. 1798–18 Jul. 1858. CD's sister. First child of Robert Waring Darwin. Wife of Henry Parker (1824). Four sons, one daughter. When widow Marianne died, her grown children were adopted by her sister, Susan Elizabeth Darwin.

Darwin, Martha Haskins "Maud" née du Puy. *See* Martha Haskins du Puy.

Darwin, Mary Eleanor. 23 Sept. 1842–16 Oct. 1842. Third child, second daughter of CD and ED. Lived only 23 days. Born and died at Down House.

Darwin, Robert Waring. 30 May 1766–13 Nov. 1848. Physician. Father of CD. Son of Erasmus Darwin I. Husband of Susannah Wedgwood (1796). Two sons, four daughters. M.D., Leiden, 1785. FRS, 1788.

Darwin, Ruth Frances. 2 Aug. 1883–1973. CD's granddaughter. Second child of Sir Horace Darwin. Wife of W. Rees Thomas.

Darwin, Sara née Sedgwick. *See* Sara Sedgwick.

Darwin, Susan Elizabeth. 3 Oct. 1803–3 Oct. 1866. CD's sister. Third child of Robert Waring Darwin. Not married. Lived at the Mount, Shrewsbury, with her father and took care of him. Adopted grown children of her widowed sister, Marianne Darwin Parker.

Darwin, William Erasmus. 27 Dec. 1839–1 Sept. 1914. Banker. First child of CD and ED. B.A., Cambridge, 1862. Husband of Sara Sedgwick (1877). No children. Was subject of CD's observations on child development published in *Mind* (1877). Managed CD's financial affairs. Amateur photographer.

Darwin, William Robert. 1894–1970. Stockbroker. CD's grandson. Fourth child of Sir George Howard Darwin. Husband of Monica Slingsby (1926).

de Bary, Heinrich Anton. 26 Jan. 1831–19 Jan. 1888. Mycologist. Professor of botany, Strasbourg, Germany. Supplied specimens to CD. Coined the word "symbiosis."

de Vries, Hugo. 16 Feb. 1848–21 May 1935. Dutch plant physiologist. Visited CD, summer 1877. Rediscoverer (in 1900) of Mendel's work on genetics. Developed mutation theory of evolution. Demonstrated that new characters can appear suddenly and are inheritable. Considered that mutations were source of genetic variation.

Dew-Smith, Albert George. 27 Oct. 1848–1903. Physiologist, Trinity College, Cambridge. Friend and business partner of Horace Darwin in Cambridge Scientific Instrument Company. Personal friend, invited to funeral of CD.

Dickens, Charles. 7 Feb. 1812–9 Jun. 1870. Novelist. Acquaintance of CD. Dickens and CD frequented the Athenaeum Club, Pall Mall, London. Author of *A Christmas Carol* (1843) and *Great Expectations* (1860–1861), among many other well-known works.

du Puy, Martha Haskins "Maud." 1861–1947. Wife of Sir George Howard Darwin, married in 1884. Four children.

Eliot, George (aka Mary Ann Cross, Marian Evans). 22 Nov. 1819–22 Dec. 1880. Novelist. Wrote under pen name George Eliot, beginning in 1857. Common-law wife (1854–1878) of George Henry Lewes until his death. Married J. W. Cross, New York banker, in 1880. Friend of Richard Litchfield, acquaintance of CD. Attended parties with CD. Author of *Silas Marner* (1861), *Romola* (1862–62), and *Middlemarch* (1874), generally considered her finest work.

Eyre, Edward John. 5 Aug. 1815–30 Nov. 1901. English explorer. Discovered large, shallow salt lake in central South Australia, 1840, which was named for him. Governor of Jamaica, 1864. Brutally suppressed black rebellion, 1865. CD supported attempt to prosecute Eyre for murder, 1866.

Farrer, Emma Cecilia "Ida." 1854–1946. CD's daughter-in-law. Wife of Sir Horace Darwin (1880). Daughter of Thomas and Frances Farrer.

Farrer, Frances "Fanny" née Erskine. First wife of Thomas Henry Farrer. Mother of Emma Cecilia "Ida" Farrer (Sir Horace Darwin's wife).

Farrer, Thomas Henry. 1833–1884. Barrister, civil servant. First wife, Frances Erskine (three sons, one daughter). Second wife, Katherine Euphemia Wedgwood. Excavated Roman ruins on his property.

Fisher, Florence Henrietta. 1864–1920. Playwright. Widow of Frederic William Maitland. Third wife of Sir Francis Darwin, married in 1913. Author of *Six Plays* (1921).

Fisher, Sir Ronald Aylmer. 17 Feb. 1890–29 Jul. 1962. Population geneticist, biomathematician, eugenicist. Protégé of Leonard Darwin. Mathematically analyzed interrelationships of mutation rates, population size, selection value, and other factors in evolutionary variation. Author of *The Genetical Theory of Natural Selection* (1930). Knighted, 1952.

FitzRoy, Robert. 5 July 1805–30 Apr. 1865. Naval officer. Royal Navy hydrographer and meteorologist. Captain of H.M.S. *Beagle,* selected CD as his gentleman-companion on voyage around the world, 1831–1836. Superb seaman. Mercurial, depressive temperament. Author of *Narrative of the Surveying Voyages of the Adventure and Beagle* (1839). MP for Durham, 1841–1843. Governor general of New Zealand, 1843–1845. FRS, 1851. Chief of Meteorological Department, Board of Trade, 1854–1865. Visited CD at Down House, 1857. Denounced evolution. Rear Admiral, 1857. Vice Admiral, 1863. Committed suicide.

Fox, William Darwin. 1805–1880. Clergyman, naturalist. CD's second cousin and closest friend at Cambridge. Retained lifelong correspondence with CD. Stimulated CD's interest in beetle collecting. Vicar of Delamere, Cheshire, 1838–1873. Retired to Sandown, Isle of Wight.

Fraser, Elizabeth Frances. 1846–1898. First wife of Leonard Darwin, married in 1882.

Galton, Sir Francis. 16 Feb. 1822–17 Jan. 1911. Statistician, eugenicist. CD's half first cousin. Founded eugenics movement. FRS, 1860. Knighted, 1909. Author of *Hereditary Genius* (1869).

Gladstone, William Ewart. 29 Dec. 1809–19 May 1898. Statesman. Four times Prime Minister, 1868–1874, 1880–1885, 1886, 1892–1894. Considered greatest British statesman of nineteenth century. Visited Down House with T. H. Huxley, 1876. Arranged Civil List pension for A. R. Wallace at CD's request, 1880. FRS, 1881.

Gordon, Charles George. 28 Jan. 1833–26 Jan. 1885. English soldier. Served in China as commander (aka Chinese Gordon) of band of Chinese fighters, 1863–1864. Governor of equatorial provinces of central Africa, fought against slave trade, 1874–1876. Governor general of Sudan, 1877–1879. Sent to Khartoum to evacuate 2,000 Egyptian civilians and 600 soldiers from assault by the Mahdi, withstood siege of 317 days, speared and beheaded in final assault, 1884.

Gray, Asa. 18 Nov. 1810–30 Jan. 1888. Botanist. Professor, Harvard University, 1842–1872. First met CD at Hunterian Museum, London, 1839. CD's strongest supporter in America. Visited and stayed at Down House, Oct. 24, 1868. Lifelong correspondent and friend of CD. CD dedicated *The Different Forms of Flowers* to "Professor Asa Gray... As a Small Tribute of Respect and Affection." Author of *Manual of the Botany of the Northern United States...*, aka *Gray's Manual* (1848), and *Darwiniana* (1876).

Gully, James Manby. 13 Mar. 1808–27 Mar. 1883. Physician, hydrotherapist. Treated many famous patients at his hydropathic spa. CD underwent cold water treatment at the Lodge, Malvern, 1849. Annie Darwin died there, 1851.

Harding, Elizabeth "Bessy." Nursery maid at Down House, 1842.

Henslow, John Stevens. 6 Feb. 1796–16 May 1861. Botanist, mineralogist, clergyman. FRS, 1818. Professor, Cambridge University, 1822–1861. CD's most influential professor there. Arranged for CD to go on H.M.S. *Beagle* voyage and received specimens sent by CD from voyage. Promoted CD's accomplishments while CD was still at sea. Described some of CD's plant specimens. Personal friend of CD. Visited Down House, 1854.

Holland, Sir Henry. 27 Oct. 1788–27 Oct. 1873. Physician. FRS, 1816. Office in London. Physician to Queen Victoria. Distant cousin to Darwins and Wedgwoods. Treated CD and other members of Darwin family at Down House.

Hooker, Sir Joseph Dalton. 30 Jun. 1817–10 Dec. 1911. Botanist. CD's closest friend and confidant. Assistant surgeon on Antarctic expedition of James Clark Ross, 1838–1843, published botanical results. Collected plants in Himalayas, 1848–1850. Specialist in plant taxonomy and plant geography. FRS, 1847. President, Royal Society, 1873–1878. Knighted, 1878, as Knight Commander of the Order of the Star of India (an honor limited to 60 individuals, reflecting service under hardship). Son of Sir William Hooker, whom he succeeded as director of Royal Botanical Gardens (Kew Gardens), 1865–1885. Supplied many botanical specimens to CD from Kew Gardens. CD and Hooker exchanged about 1,400 letters. Helped C. Lyell organize the joint reading of CD's and Wallace's paper at Linnean Society meeting, 1858. Pallbearer at CD's funeral, 1882.

Huxley, Thomas Henry. 4 May 1825–29 June 1895. Zoologist, comparative anatomist. Earned nickname of "Darwin's Bulldog" for his staunch defense of CD at Oxford debate with Bishop Samuel Wilberforce in 1860 and in published articles. Surgeon on H.M.S. *Rattlesnake*, surveying eastern coast of Australia, 1846–1850. Studied marine invertebrates. FRS, 1851. Lecturer/professor, Royal School of Mines, 1854–1884. Hunterian Professor, Royal College of Surgeons, 1862–1869. Fullerian Professor of Physiology, Royal Institution,

1855–1858, 1866–1869. Close personal friend of CD since 1855. Frequent visitor to Down House. Pallbearer at CD's funeral, 1882. President, Royal Society, 1883–1885. Author of *Evidence as to Man's Place in Nature* (1863), and *Collected Essays* (9 vols., 1893–1894).

Innes, John Brodie. 1817–1894. Clergyman. Vicar of Downe, 1846–1862. Friend of CD. Left Downe in 1862 after inheriting property in Scotland. Personal friend, invited to funeral of CD.

Jenyns, Leonard. 25 May 1800–1 Sept. 1893. Anglican priest, naturalist, ichthyologist. Fellow beetle collector and friend of CD at Cambridge. Brother-in-law of Professor J. S. Henslow. Declined offer to sail with Capt. FitzRoy as naturalist on H.M.S. *Beagle*, thus paving the way for CD. CD asked him to describe fish specimens collected during *Beagle* voyage. Author of *Memoir of the Reverend John Stevens Henslow* (1862). Adopted the name Blomefield, 1871.

Keith, Sir Arthur. 5 Feb. 1866–7 Jan. 1955. Surgeon. FRS, 1913. Knighted, 1921. Heavily involved in purchase of Down House for British Association for the Advancement of Science and its later acquisition by Royal College of Surgeons.

Kelvin, Lord. *See* William Thomson.

Keynes, Sir Geoffrey. 25 Mar. 1887–5 July 1982. Physician and surgeon. Knighted, 1955. Husband of Margaret Elizabeth Darwin (1917). Brother of John Maynard Keynes. Father of Richard Darwin Keynes.

Keynes, John Maynard. 5 June 1833–21 Apr. 1946. Economist. Brother-in-law of Margaret Elizabeth née Darwin Keynes. Originator of "New Economics," with his deficit-spending ideas still widely held. Advised government spending to counter deflation and depression, which became President Franklin D. Roosevelt's plan for recovery from the Great Depression.

Keynes, Margaret Elizabeth née Darwin. *See* Margaret Elizabeth Darwin.

Keynes, Richard Darwin. 14 Aug. 1919–12 June 2010. Physiologist. Great grandson of CD. Son of Geoffrey Keynes and Margaret Elizabeth Darwin. Nephew of John Maynard Keynes. Married Anne Pinsent Adrian in 1945. Father of 4 sons, including Randal Keynes, the author of *Annie's Box*. FRS, 1959. Author of 4 major Darwin books: *The* Beagle *Record* (1979), *Charles Darwin's* Beagle *Diary* (1988), *Charles Darwin's Zoology Notes and Specimen Lists from H.M.S.* Beagle (2000), and *Fossils, Finches, and Fuegians* (2002).

Lady Hope. *See* Elizabeth Reid Cotton.

Lane, Edward Wickstead. 1823–1889. Physician. Owner of hydropathic establishment at Moor Park, Surrey. CD sought treatment there, beginning in 1857. Personal friend, invited to funeral of CD.

Lettington, Henry. 1822/1823–?. Gardener at Down House, 1854–1879. Assisted CD with botanical experiments. On list of personal friends, invited to CD's funeral.

Lewes, George Henry. 1817–1878. Scientific writer and critic. Common-law husband (1854–1878) of Mary Ann Cross, aka George Eliot.

Litchfield, Henrietta Emma née Darwin. *See* Henrietta Emma Darwin.

Litchfield, Richard Buckley. 6 Jan. 1832–11 Jan. 1903. Lawyer. Son-in-law of CD. Husband of Henrietta Darwin, married in 1871. Founded Working Men's College.

Longfellow, Henry Wadsworth. 27 Feb. 1807–24 Mar. 1882. American poet. Visited CD on Isle of Wight, 1868.

Lowell, James Russell. 22 Feb. 1819–12 Aug. 1891. American literary critic, poet, and diplomat. First editor of *Atlantic Monthly*, 1857. American Minister Plenipotentiary to the Court of St. James, London, 1880–1885. Pallbearer at CD's funeral.

Lubbock, Sir John William, 3rd Baronet. 26 Mar. 1803–20 June 1865. Astronomer, mathematician, banker. Owner of large estate and CD's neighbor at Downe. Sold CD the land for his walking path. FRS, 1829. Treasurer and vice president, Royal Society, 1830–1835, 1838–1845. First vice-chancellor, London University, 1837–1842.

Lubbock, Sir John, 4th Baronet. 30 Apr. 1834–28 May 1913. Banker, politician, naturalist. Son of Sir John Lubbock, 3rd Baronet. Neighbor of CD until 1861. CD, considering him a member of the family, trained his young neighbor in entomology and natural history. Strong supporter of CD and natural selection. Liberal MP, 1870, 1874. Became spokesman for science. FRS, 1858. Suggested CD be given Westminster Abbey funeral and helped organize this event. Pallbearer at CD's funeral, 1882. First Baron Avebury, 1900.

Ludwig, Camilla. Governess at Down House, 1859–1865.

Lyell, Sir Charles. 14 Nov. 1797– 22 Feb. 1875. Geologist, lawyer. Geological mentor of CD, his close friend, correspondent, and supporter. Visited CD at Down House. Professor of geology, King's College, London, 1831–1833. President, Geological Society, 1834–1836, 1849–1850. President, British Association for the Advancement of Science, 1864. FRS, 1826. Knighted, 1848. Copley Medal, 1858. Author of *Principles of Geology* (3 vols., 1831–1833), which CD read during voyage of H.M.S. *Beagle*. Promoted uniformitarian geology. Paved the way for CD's acceptance in scientific society. Orchestrated joint reading of CD's and Wallace's paper at Linnean Society meeting, 1858. Traveled widely and published accounts of his visits to the United States. Author of *Elements of Geology* (1838) and *The Geological Evidence of the Antiquity of Man* (1863).

Malthus, Thomas Robert. 13 Feb. 1766–23 Dec. 1834. Clergyman, political economist. Quantified relationship between population growth and food supply in *Essay on the Principle of Population* (1798), which stimulated CD in 1838 (and A. R. Wallace 20 years later) to develop concept of natural selection. FRS, 1819.

Maitland, Frederic William. 1850–1906. Professor of laws of England, Cambridge, 1888–1906. First husband of Florence Henrietta Fisher (1886), who later became third wife of Sir Francis Darwin, married in 1913.

Martineau, Harriet. 12 June 1802–27 June 1876. Feminist, social reformer, traveler. Writer on religion, economics, government. Close friend of Erasmus Alvey Darwin, acquaintance of CD. Author of *Poor Laws and Paupers Illustrated* (1833), *Society in America* (1837), *Illustrations of Taxation* (1843), and *Autobiography* (1877).

Massingberd, Charlotte Mildred. 1868–1940. CD's daughter-in-law. Second wife of Leonard Darwin, married in 1900.

Mendel, (Johann) Gregor. 22 July 1822–6 Jan. 1884. Augustinian friar at Brno, Moravia. Abbot, 1868. Founder of modern genetics. Experiments with pea plants illuminated first two laws of inheritance: law of segregation, and law of independent assortment. Published results in German in *Proceedings of the Natural Science Society of Brünn*, 1866, which CD did not see. This paper rediscovered in 1900 by Hugo de Vries.

Mivart, St. George Jackson. 30 Nov. 1827–1 Apr. 1900. Comparative anatomist, barrister. Darwin antagonist. Lecturer in biology, St. Mary's Roman Catholic College, Kensington. Catholic convert, 1844. Alienated church and scientific community. FRS, 1869. Secretary, Linnean Society, 1874–1880. Author of *The Genesis of Species* (1871), in which he strongly criticized *On the Origin of Species* and *The Descent of Man*. Attempted to reconcile evolutionary theory and Catholicism. Excommunicated, 1900.

Moore, Sir Norman. 1847–1922. Physician, St. Bartholomew's Hospital. Attended CD in his last illness. Personal friend, invited to funeral of CD.

Murray, John. 16 Apr. 1808–2 Apr. 1892. CD's main publisher, from 1845 at 50 Albemarle Street, London. Personal friend, invited to funeral of CD.

Nightingale, Florence. 12 May 1820–13 Aug. 1910. Nurse, hospital reformer. Volunteered her services in Crimean War and set up hospital in Scutari, 1854. Emphasized stringent sanitation and reduced death rate from cholera, typhus, and dysentery from 50 percent to 2 percent. Founded Nightingale Home at St. Thomas's Hospital for nurses' training. First woman to receive Order of Merit, 1907.

Norton, Charles Eliot. 1827–1908. Professor of art history, Harvard University. Brother-in-law of Sara Sedgwick, wife of William Alvey Darwin, married in 1868. Visited CD at Down House.

Ouless, Walter William. 1848–1933. Painter. Created oil portrait of CD, 1875.

Owen, Sir Richard. 20 July 1804–18 Dec. 1892. Comparative anatomist. CD selected him to describe fossil mammals collected during H.M.S. *Beagle* voyage. FRS, 1834. Hunterian Professor, Royal College of Surgeons, 1836–1856. Superintendent of natural history departments, British Museum, 1856–1884. President, British Association for the Advancement of Science, 1858. Major force in establishing Natural History Museum, South Kensington, and became its first director, 1881. Knighted, 1884. Initial friendship with CD turned to hatred with publication of *On the Origin of Species*. Became bitter, envious critic of CD, and was loathed by CD. Anonymously published spiteful review of *The Origin* in *Edinburgh Review* (1860).

Paley, William. July 1743–25 May 1805. Clergyman. Tutor at Christ's College, Cambridge, 1771–1774. CD allegedly occupied Paley's same rooms at Cambridge, years later. CD admired Paley's *Natural Theology* (1802), but eventually rejected his assumption that natural forces could not be responsible for organized complexity.

Palmerston, Lord (aka Henry John Temple, 3rd Viscount Palmerston). 20 Oct. 1784–18 Oct. 1865. Influential Whig-

Liberal statesman, with 30 years of service as British foreign secretary or as Prime Minister (1855–1858, 1859–1865). Nominated CD for knighthood in 1859, but was thwarted by Bishop of Oxford, Samuel Wilberforce.

Parker, Henry. 1788–1856. Physician, surgeon. CD's brother-in-law. Husband of Marianne Darwin, married in 1824. Four sons, one daughter. After Marianne's death, their grown children were adopted by her sister, Susan Elizabeth Darwin.

Parker, Marianne née Darwin. *See* Marianne Darwin.

Parslow, Eliza. ?–1881. ED's personal maid. Married Joseph Parslow, butler at Down House. Had dressmaking school in Downe.

Parslow, Joseph. 1809/1810–1898. CD's manservant at 12 Upper Gower Street, circa 1840. Butler at Down House until 1875. Considered part of Darwin family. Referred to as "aged Parslow" in Dickens's *Great Expectations*. Darwin family referred to him as "venerable Parslow." After CD's death, received pension and rent for a house. Walked in CD's funeral procession.

Rajon, Paul. 1842/1843–1888. Engraver. Engraved Ouless's oil portrait of CD on copper, 1875.

Raverat, Gwendolen Mary née Darwin. *See* Gwendolen Mary Darwin.

Raverat, Jacques. 1885–1925. Artist. Husband of Gwendolen Mary Darwin (1911).

Reed, George Varenne. 1816–1886. Anglican clergyman. Tutor to George, Francis, Leonard, and Horace Darwin before they went to Clapham Grammar School. Personal friend, invited to funeral of CD.

Romanes, George John. 2 May 1848–23 May 1894. Zoologist, comparative psychologist. Close personal friend and protégé of CD, 1874 on. FRS, 1879. Frequent correspondent and visitor at Down House. Personal friend, invited to funeral of CD. Author of *Animal Intelligence* (1882), *Mental Evolution* *in Animals* (1883), and *Darwin and after Darwin* (3 vols., 1892–1897).

Ruck, Amy Richenda. 1850–1876. CD's daughter-in-law. First wife of Sir Francis Darwin, married in 1874. Mother of Bernard Darwin. Died in childbirth.

Ruck, Mary. ?–? Mother of Amy Ruck. Became friend of ED.

Sach, Julius von. 2 Oct. 1832–29 May 1897. German botanist. Professor at Würzburg. Critic of CD's and Francis Darwin's botanical work. Francis studied latest methods in emerging science of plant physiology in Sach's laboratory.

Sedgwick, Sara. 1839–1902. CD's daughter-in-law. Wife of William Erasmus Darwin, married in 1877. From Cambridge, MA. No children.

Sowerby, George Brettingham II. 1812–1844. Drew illustrations for all of CD's barnacles, 1851–1854. Was at Down House to draw orchids for *Fertilisation of Orchids*, 1861.

Stokes, John Lort. 1812–1885. Naval officer, draughtsman. Mate and assistant surveyor on second voyage of H.M.S. *Beagle*, 1831–1836. Shared poop cabin with CD. On third voyage of *Beagle*, Stokes named Darwin Harbour after his shipmate and friend, 1839. Admiral, 1877. Author of *Discoveries in Australia* (2 vols., 1846).

Sulivan, Bartholomew James. 18 Nov. 1810–1 Jan. 1890. Naval officer, hydrographer. Second lieutenant on H.M.S. *Beagle* with CD, 1831–1836. Knighted, 1869. Admiral, 1877. Visited CD at Down House with John Wickham (another *Beagle* officer), 1861.

Tennyson, (Lord) Alfred. 6 Aug. 1809–6 Oct. 1892. Poet. Tennyson called on CD several times during their stay at Freshwater, on the Isle of Wight. Author of *In Memoriam and Other Poems* (1833–1850). Poet Laureate, 1850.

Thomas, Ruth Frances née Darwin. *See* Ruth Frances Darwin.

Thompson, William (Lord Kelvin). 26 June 1824–7 Dec. 1907. Physicist. Professor of natural philosophy, University of Glasgow, 1849–1899. FRS, 1851. Knighted, 1866. President, Royal Society, 1890. Most prominent astronomical physicist of the time. Expert in thermodynamics. Kelvin temperature scale named for him. Came into conflict with CD over age of earth, with Thompson's estimate of its age being much too young. George Darwin became his protégé.

Thorley, Catherine. Governess at Down House, 1850–1856. Was with Annie Darwin when she died at Malvern, 1851. Personal friend, invited to funeral of CD.

Tyndall, John. 2 Aug. 1820–4 Dec. 1893. Physicist. FRS, 1852. Superintendent, Royal Institution, 1867–1887. Popularizer of science. Close friend of T. H. Huxley. Stayed overnight at Down House with Asa Gray and J. D. Hooker, 1868. Personal friend, invited to funeral of CD.

Vines, Sydney Howard. 31 Dec. 1849–4 Apr. 1934. Botanist. FRS, 1885. Reader in botany, Cambridge University. Colleague of Francis Darwin.

Wallace, Alfred Russel. 8 Jan. 1823–7 Nov. 1913. Naturalist, traveler, collector. Independent codiscoverer of evolution by natural selection, but deferred to CD. Royal Medal, 1868. First recipient of Darwin Medal, 1890. FRS, 1905. Order of Merit, 1908. Copley Medal, 1908. First recipient of Darwin-Wallace Medal of Linnean Society, 1908. Collector of specimens in Amazon, 1848–1852, and Malay Archipelago, 1854–1862. CD and J. D. Hooker raised issue of civil pension for Wallace. Often visited CD at Down House. Wallace's letter and paper to CD in 1858 stimulated CD to publish his ideas on evolution. C. Lyell and J. D. Hooker communicated CD's and Wallace's papers to Linnean Society, 1858. Pallbearer at CD's funeral, 1882. Author of *Palm Trees of the Amazon* (1853), *Travels on the Amazons* (1853), *Malay Archipelago* (1869), *Contributions to the Theory of Natural Selection* (1870), *Geographical Distribution of Animals*

(2 vols., 1876), *Island Life* (1880), *Darwinism* (1889), and *My Life* (2 vols., 1905). Best known as biogeographer. Wallace's Line (a faunal boundary line) and genus *Wallacea* named for him.

Waring, Anne. 1662–1722. CD's great-great-grandmother.

Wedgwood, Caroline Sarah née Darwin. *See* Caroline Sarah Darwin.

Wedgwood, Emma. 2 May 1808–2 Oct. 1896. CD's wife (and first cousin), married in 1839. Ninth and last child of Josiah Wedgwood II. Looked after and nursed CD their entire married life. Center of Darwin family. Devoutly religious Unitarian. Buried in Downe churchyard.

Wedgwood, Frances "Fanny" née Mackintosh. 1800–1889. CD's sister-in-law. Wife of Hensleigh Wedgwood, married in 1832.

Wedgwood, Frances "Fanny." 1806–20 Aug. 1832. CD's first cousin and sister-in-law. ED's sister. Eighth child of Josiah Wedgwood II.

Wedgwood, Hensleigh. 22 Jan. 1803–2 June 1891. Barrister, philologist. CD's first cousin and brother-in-law. ED's brother. Seventh child of Josiah Wedgwood II. Husband of Fanny Mackintosh (1832). Three sons, three daughters. Police magistrate, 1831–1837. Registrar of hackney cabs, 1838–1849. Author of *A Dictionary of English Etymology* (3 vols., 1859–1865), and *The Origin of Language* (1866).

Wedgwood, Josiah I. 12 July 1730–3 Jan. 1795. Industrialist. Master potter, founder of Josiah Wedgwood & Sons, Eturia, Staffordshire. CD's maternal grandfather. Father of CD's mother Susannah. Close friend of Erasmus Darwin. FRS, 1783.

Wedgwood, Josiah II. 1769–12 July 1843. Master potter and partner in Josiah Wedgwood & Sons, 1795–1841. Lived at Maer Hall, Staffordshire. CD's maternal "Uncle Jos." ED's father. Husband of Elizabeth "Bessy" Allen (1792). Four sons, five daughters. Persuaded CD's father to allow him to accept trip around world on H.M.S. *Beagle*. Whig MP for Stoke-on-Trent, 1832–1834.

Wedgwood, Josiah III. 1795–11 Mar. 1880. Master potter, senior partner in Josiah Wedgwood & Sons, 1841–1844. CD's first cousin and brother-in-law. ED's brother. Husband of Caroline Sarah Darwin, married in 1837. Four daughters. Lived at Leith Hill Place, Surrey, 1844.

Wedgwood, Katherine Euphemia. 1839–1934. Fourth child of Hensleigh and Fanny Wedgwood. Wife of Sir Thomas Farrer (1873). CD's first cousin once removed. Stayed at Down House.

Wedgwood, Sarah Elizabeth "Bessy." 1793–1800. First child of Josiah Wedgwood II. CD's first cousin. ED's unmarried sister.

Wedgwood, Thomas. 14 May 1771–10 Jul. 1805. CD's uncle. Unmarried fourth child of Josiah Wedgwood I. Invented chemical photographic process, 1807. Called "the first photographer." Published research on heat and light.

Wharton, Henry James. 1798–1859 Clergyman. Headmaster of William Erasmus Darwin's preparatory school, 1850.

REFERENCES

Anderson, William D., Jr. 2002. Andrew C. Moore's "Evolution Once More": The Evolution-Creationism Controversy from an Early 1920s Perspective. *Bulletin of the Alabama Museum of Natural History* 22: 1–35.

Atkins, Hedley. 1974. *Down: The Home of the Darwins*. Royal College of Surgeons of England. London.

Atran, Scott and Jeremy Ginges. 2012. Religious and Sacred Imperatives in Human Conflict. *Science* 336: 855–857.

Ayres, Peter. 2008. *The Aliveness of Plants: The Darwins at the Dawn of Plant Science*. Pickering & Chatto, London.

Bailes, E., F. Gao, F. Bibollet-Ruche, V. Courgnaud, M. Peeters, P. Marx, B. H. Han, and P. M. Sharp. 2003. Hybrid Origin of SIV in Chimpanzees. *Science* 300: 1713.

Barkow, J. H., L. Cosmides, and J. Tooby (Eds.). 1992. *The Adapted Mind*. Oxford University Press, Oxford.

Barlow, Nora (Ed.). 1933. *Charles Darwin's Diary of the Voyage of the H.M.S. Beagle*. Cambridge University Press, Cambridge.

Barlow, Nora (Ed.) 1945. *Charles Darwin and the Voyage of the Beagle*. Pilot Press, London.

Barlow, Nora (Ed.) 1958. *The Autobiography of Charles Darwin, 1809–1882: With Original Omissions Restored*. W. W. Norton, New York.

Barlow, Nora (Ed.) 1967. *Darwin and Henslow: The Growth of an Idea.* John Murray, London.

Barrett, Paul, Peter J. Gautrey, Sandra Herbert, David Kohn, and Sydney Smith (Eds.). 1987. *Charles Darwin's Notebooks, 1836–1844.* Cornell University Press, Ithaca, NY.

Bartlett, John. 1968. *Bartlett's Familiar Quotations: A Collection of Passages, Phrases, and Proverbs Traced to their Sources in Ancient and Modern Literature.* 14th edition. Emily Morison Beck (Ed.). Little, Brown, Boston.

Beers, M. H. and R. Berkow. 1999. *The Merck Manual of Diagnosis and Therapy.* 17th edition. Merck Research Laboratories, Whitehouse Station, NJ.

Benét, William Rose. 1965. *The Reader's Encyclopedia.* 2nd edition. 2 vols. Thomas Y. Crowell, New York.

Bennett, J. H. 1983. Introduction. In *Natural Selection, Heredity, and Eugenics.* J. H. Bennett (Ed.), 1–50. Clarendon Press, Oxford.

Berra, Tim M. 1990. *Evolution and the Myth of Creationism.* Stanford University Press, Stanford, CA.

Berra, Tim M. 1998. *A Natural History of Australia.* University of New South Wales Press/Academic Press, Sydney/San Diego.

Berra, Tim M. 2008. Charles Darwin's paradigm shift. *The Beagle, Records of the Museum and Art Galleries of the Northern Territory* 24: 1–5.

Berra, Tim M. 2009. *Charles Darwin: The Concise Story of an Extraordinary Man.* Johns Hopkins University Press, Baltimore.

Berra, Tim M. 2013. Wallace's Acceptance of Darwin's Priority in His Own Words. *Linnean* 29: in press.

Berra, Tim M., Gonzalo Alvarez, and Francisco C. Ceballos. 2010a. Was the Darwin/Wedgwood Dynasty Adversely Affected by Consanguinity? *BioScience* 60: 376–383.

Berra, Tim M., Gonzalo Alvarez, and Kate Shannon. 2010b. The Galton-Darwin-Wedgwood Pedigree of H. H. Laughlin. *Biological Journal of the Linnean Society* 101: 228–241.

Bloch, Maurice. 2008. Why Religion Is Nothing Special but Is Central. *Philosophical Transactions of the Royal Society, B* 363: 2055–2061.

Bowlby, John. 1990. *Charles Darwin: A New Life.* Norton, New York.

Bowler, Peter J. 1984. *Evolution: the History of an Idea.* University of California Press, Berkeley.

Brodribb, Tim J. and Scott A. M. McAdam. 2011. Passive Origins of Stomatal Control in Vascular Plants. *Science* 331: 582–585.

Brown, E[rnest]. W. 1916. The Scientific Work of Sir George Darwin. In Darwin, George H. *Scientific Papers by Sir George Howard Darwin.* Vol. 5, *Supplemental Volume*, F. J. M. Stratton and J. Jackson (Eds.), xxxiv–lv. Cambridge University Press, Cambridge.

Brown, Kyle S., et al. 2012. An Early and Enduring Advanced Technology Originating 71,000 Years Ago in South Africa. *Nature* 491: 590–593.

Browne, Janet. 1978. The Charles Darwin–Joseph Hooker Correspondence: An Analysis of Manuscript Resources and Their Use in Biography. *Journal of the Society for Bibliography of Natural History* 8: 351–366.

Browne, Janet. 1995. *Charles Darwin: A Biography. Vol. 1, Voyaging.* Princeton University Press, Princeton, NJ.

Browne, Janet. 2002. *Charles Darwin: A Biography. Vol. 2, The Power of Place.* Knopf, New York.

Browne, Janet. 2003. Charles Darwin as a Celebrity. *Science in Context* 16: 175–194.

Browne, Janet. 2010. Darwin's Intellectual Development: Biography, History and Commemoration. In *Darwin.* William Brown and Andrew C. Fabian (Eds.), 1–30. Cambridge University Press, Cambridge.

Burchfield, J. D. 1975. *Lord Kelvin and the Age of the Earth.* Science History, New York.

Burkhardt, Frederick, et al. (Eds.). 1985–. *The Correspondence of Charles Darwin.* 19+ vols. Cambridge University Press, Cambridge. [This project is expected to include 30 volumes and to be completed around 2025.]

Carroll, Sean B. 2005. *Endless Forms Most Beautiful: The New Science of Evo Devo and the Making of the Animal Kingdom.* W. W. Norton, New York.

Cattermole, M. J. G. and A. F. Wolfe. 1987. *Horace Darwin's Shop: A History of the Cambridge Scientific Instrument Company, 1878–1968.* Adam Hilger, Bristol and Boston.

Choy, C. M., C. W. Lam, L. T. Cheung, C. M. Briton-Jones, L. P. Cheung, and C. J. Haines. 2002. Infertility, Blood Mercury Concentrations and Dietary Seafood Consumption: A Case-

Control Study. *British Journal of Obstetrics and Gynaecology* 109: 1121–1125.
Colp, Ralph, Jr. 2008. *Darwin's Illness*. University Press of Florida, Gainesville.
Costa, James T. (Ed.). 2009. *The Annotated Origin: A Facsimile of the First Edition of "On the Origin of Species" by Charles Darwin*. Harvard University Press, Cambridge, MA.
Dalrymple, G. Brent. 1991. *The Age of the Earth*. Stanford University Press, Stanford, CA.
Darwin, Bernard. 1955. *The World that Fred Made: An Autobiography*. Chatto and Windus, London.
Darwin, Charles (Ed.). 1838–1843. *The Zoology of the Voyage of the H.M.S. Beagle, under the Command of Captain FitzRoy... during the Years 1832–1836*. 5 parts. Smith Elder, London.
Darwin, Charles. 1839. *Journal of Researches into the Geology and Natural History of Various Countries Visited by H.M.S. Beagle, under the Command of Capt. Fitzroy, R.N.* Henry Colburn, London.
Darwin, Charles. 1842. *The Structure and Distribution of Coral Reefs: Being the First Part of the Geology of the Voyage of the Beagle, under the Command of Capt. Fitzroy, R.N. during the Years 1832–1836*. Smith Elder, London.
Darwin, Charles. 1844. *Geological Observations on the Volcanic Islands Visited during the Voyage of H.M.S. Beagle, Together with Some Brief Notices of the Geology of Australia and the Cape of Good Hope: Being the Second Part of the Geology of the Beagle, under the Command of Capt. Fitzroy, R.N. during the Years 1832–1836*. Smith Elder, London.
Darwin, Charles. 1846. *Geological Observations on South America: Being the Third Part of the Geology of the Beagle, under the Command of Capt. Fitzroy, R.N. during the Years 1832–1836*. Smith Elder, London.
Darwin, Charles. 1851a. *A Monograph of the Fossil Lepadidae; or, Pedunculated Cirripedes of Great Britain*. Vol. 1. Palaeontographical Society, London.
Darwin, Charles. 1851b. *A Monograph of the Sub-Class Cirripedia, with Figures of All the Species*. Vol. 1, *The Lepadidae; or, Pedunculated Cirripedes*. Ray Society, London.
Darwin, Charles. 1854a. *The Balanidae (or Sessile Cirripedes), etc.* Vol. 2. Ray Society, London.

Darwin, Charles. 1854b. *A Monograph of the Fossil Balanidae and Verrucidae of Great Britain*. Vol. 2. Palaeontographical Society, London.

Darwin, Charles. 1859. *On the Origin of Species by Means of Natural Selection; or, The Preservation of Favoured Races in the Struggle for Life*. John Murray, London.

Darwin, Charles. 1862. *On the Various Contrivances by which British and Foreign Orchids Are Fertilized by Insects and on the Good Effects of Intercrossing*. John Murray, London.

Darwin, Charles. 1865. On the Movements and Habits of Climbing Plants. *Journal of the Proceedings of the Linnean Society of London* 9: 1–128. Published in book form by John Murray, London, 1875.

Darwin, Charles. 1868. *The Variation of Animals and Plants under Domestication*. 2 vols. John Murray, London.

Darwin, Charles. 1870–1871. *The Descent of Man and Selection in Relation to Sex*. 2 vols. John Murray, London.

Darwin, Charles. 1872. *The Expression of Emotions in Man and Animals*. John Murray, London.

Darwin, Charles. 1875. *Insectivorous Plants*. John Murray, London.

Darwin, Charles. 1876. *The Effects of Cross- and Self-Fertilisation in the Vegetable Kingdom*. John Murray, London.

Darwin, Charles. 1877a. A Biographical Sketch of an Infant. *Mind: A Quarterly Review of Psychology and Philosophy* 2: 285–294.

Darwin, Charles. 1877b. *The Different Forms of Flowers on Plants of the Same Species*. John Murray, London.

Darwin, Charles. 1879. Preliminary Notice. In Ernst Krause, *Erasmus Darwin*. Translated from the German by W. S. Dallas, 1–127. John Murray, London.

Darwin, Charles. 1880. Assisted by Francis Darwin. *The Power of Movement in Plants*. John Murray, London.

Darwin, Charles. 1881. *The Formation of Vegetable Mould, through the Action of Worms, with Observations on Their Habits*. John Murray, London.

Darwin, Charles and Alfred Wallace. 1858. On the Tendency of Species to Form Varieties; and On the Perpetuation of Varieties and Species by Natural Means of Selection. *Journal of the Proceedings of the Linnean Society (Zoology)* 3: 45–62.

Darwin, Francis. 1876. On the Primary Vascular Dilatation in Acute Inflammation. *Journal of Anatomy and Physiology* 10: 1–16.

Darwin, Francis (Ed.). 1887. *The Life and Letters of Charles Darwin, including an Autobiographical Chapter*. 3 vols. John Murray, London.
Darwin, Francis. 1895. *Elements of Botany*. Cambridge University Press, Cambridge.
Darwin, Francis. 1898. Observations on Stomata. *Philosophical Transactions of the Royal Society of London, B* 190: 561–621.
Darwin, Francis (Ed.). 1909. *The Foundations of "The Origin of Species": Two Essays Written in 1842 and 1844 by Charles Darwin*. Cambridge University Press, Cambridge.
Darwin, Francis. 1914. [Obituary] William Erasmus Darwin. *Christ's College Magazine* 29: 16–23.
Darwin, Francis. 1916. Memoir of Sir George Darwin. In Darwin, George H. *Scientific Papers by Sir George Howard Darwin*, Vol. 5, *Supplemental Volume*: ix–xxxiii. Cambridge University Press, Cambridge.
Darwin, Francis. 1917. *Rustic Sounds and Other Studies in Literature and Natural History*. John Murray, London.
Darwin, Francis and E. H. Acton. 1894. *Practical Physiology of Plants*. Cambridge University Press, Cambridge.
Darwin, Francis and A. C. Seward (Eds.). 1903. *More Letters of Charles Darwin*. John Murray, London.
Darwin, George H. 1873. On the Beneficial Restrictions to Liberty of Marriage. *Contemporary Review* 22: 412–426.
Darwin, George [H]. 1874. Note upon the Article "Primitive Man—Tylor and Lubbock," in No. 273. *Quarterly Review* 137, No. 274: 587–588.
Darwin, George H. 1875. Marriages Between First Cousins in England and Their Effects. *Journal of the Statistical Society* 38: 153–184.
Darwin, George H. 1877. On the Influence of Geological Changes on the Earth's Axis of Rotation. *Philosophical Transactions of the Royal Society of London* 167: 271–312.
Darwin, George H. 1879a. On the Bodily Tides of Viscous and Semi-Elastic Spheroids and on the Ocean Tides upon a Yielding Nucleus. *Philosophical Transactions of the Royal Society of London* 170: 1–35.
Darwin, George H. 1879b. On the Precession of a Viscous Spheroid, and on the Remote History of the Earth. *Philosophical Transactions of the Royal Society of London* 170: 447–538.

Darwin, George H. 1880. On the Secular Changes in the Orbit of a Satellite Revolving around a Tidally Disturbed Planet. *Philosophical Transactions of the Royal Society of London* 171: 713–891.

Darwin, George H. 1881. On the Tidal Friction of a Planet Attended by Several Satellites, and on the Evolution of the Solar System. *Philosophical Transactions of the Royal Society of London* 172: 491–535.

Darwin, George H. 1898. *The Tides and Kindred Phenomena in the Solar System*. John Murray, London.

Darwin, George H. 1907–1916. *The Scientific Papers of Sir George Darwin*. 5 vols. Cambridge University Press, Cambridge. Reissued by Cambridge University Press, 2009.

Darwin, George H. 1911. s.v. "tide." *Encyclopaedia Britannica,* 11th edition, 25: 938–961.

Darwin, Horace. 1901. On the Small Vertical Movements of a Stone Laid on the Surface of the Ground. *Proceedings of the Royal Society of London* 68: 253–261.

Darwin, Leonard. 1897. *Bimetallism: A Summary and Examination of the Arguments for and against a Bimetallic System of Currency*. John Murray, London.

Darwin, Leonard. 1899. Thirty-First Day: Major Leonard Darwin Called and Examined. In *Report of the Committee Appointed to Inquire into the Indian Currency, 1898*, Indian Currency Committee, 222–231. Her Majesty's Stationary Office, London.

Darwin, Leonard. 1900. Thomas Wedgwood. *Dictionary of National Biography* 60: 146.

Darwin, Leonard. 1903. *Municipal Trade: The Advantages and Disadvantages Resulting from the Substitution of Representative Bodies for Private Proprietors in the Management of Industrial Undertakings*. John Murray, London.

Darwin, Leonard. 1907. *Municipal Ownership: Four Lectures Delivered at Harvard, 1907*. John Murray, London.

Darwin, Leonard. 1926. *The Need for Eugenic Reform*. John Murray, London.

Darwin, Leonard. 1928. *What is Eugenics?* Watts, London.

Darwin, Leonard. 1929. Memories of Down House. *Nineteenth Century* 106: 118–123.

Darwin, L[eonard], Arthur Schuster, and E. Walter Maunder. 1889. On the Total Solar Eclipse of August 29, 1886. *Philosophical Transactions of the Royal Society, A* 180: 291–350.

Davenport-Hines, Richard. 2004. Gordon, Charles George (1833–1885). *Oxford Dictionary of National Biography*. Oxford University Press, Oxford.

Davis, Roy. 2012. How Charles Darwin Received Wallace's Ternate Paper 15 Days Earlier than He Claimed: A Comment on Van Wyhe and Rookmaaker (2012). *Biological Journal of the Linnean Society* 105: 472–477.

Dawkins, Richard. 1986. *The Blind Watchmaker*. W. W. Norton, New York.

de Waal, Frans. 1989. *Peacemaking among Primates*. Harvard University Press, Cambridge, MA.

de Waal, Frans. 2005. *Our Inner Ape*. Riverhead Books, New York.

de Waal, Frans. 2012. The Antiquity of Empathy. *Science* 336: 874–876.

de Waal, Frans. 2013. *The Bonobo and the Atheist: In Search of Humanism among the Primates*. W. W. Norton, New York.

Dennett, Daniel C. 1995. *Darwin's Dangerous Idea*. Touchstone, New York.

Dennett, Daniel C. 2006. *Breaking the Spell: Religion as a Natural Phenomenon*. Viking, New York.

Desmond, Adrian. 1997. *Huxley: From Devil's Disciple to Evolution's High Priest*. Perseus Books, Reading, MA.

Desmond, Adrian and James Moore. 1991. *Darwin: The Life of a Tormented Evolutionist*. Warner Books, New York.

Desmond, Adrian and James Moore. 2009. *Darwin's Sacred Cause: How Hatred of Slavery Shaped Darwin's Views on Human Evolution*. Houghton Mifflin Harcourt, Boston.

Dobzhansky, Theodosius. 1937. *Genetics and "The Origin of Species."* Columbia University Press, New York.

Dobzhansky, Theodosius. 1973. Nothing in Biology Makes Sense Except in the Light of Evolution. *American Biology Teacher* 35: 125–129.

Edwards, A. W. F. 2004. Darwin, Leonard (1850–1943). *Oxford Dictionary of National Biography*. Oxford University Press, Oxford.

Eugenics Review. 1943. Leonard Darwin, 1850–1943. *Eugenics Review* 34: 109–116.

Fara, Patricia. 2012. *Erasmus Darwin: Sex, Science, and Serendipity*. Oxford University Press, Oxford.

Fisher, Osmond. 1881. *Physics of the Earth's Crust*. Macmillan, London.

Fisher, Ronald A. 1918. The Correlation between Relatives on the Supposition of Mendelian Inheritance. *Philosophical Transactions of the Royal Society of Edinburgh* 52: 399–433.

Fisher, Ronald A. 1930. *The Genetical Theory of Natural Selection*. Dover, New York. Reprint, 1958.

FitzRoy, Robert. 1839. *Narrative of the Surveying Voyages of His Majesty's Ships Adventure and Beagle, between the Years 1928 and 1836, describing their Examination of the Southern Shores of South America, and the Beagle's Circumnavigation of the Globe*. 3 vols. Henry Colburn, London.

Forrest, Barbara and Paul R. Gross. 2004. *Creationism's Trojan Horse: The Wedge of Intelligent Design*. Oxford University Press, Oxford.

Freeman, R. B. 1977. *The Works of Charles Darwin: An Annotated Bibliographical Handlist*. 2nd edition, revised and enlarged. Archon Books, Hamden, CT.

Freeman, R. B. 1978. *Charles Darwin: A Companion*. Archon Books, Hamden, CT.

Freeman, R. B. 1982. The Darwin Family. In *Charles Darwin: A Commemoration*. R. J. Berry (Ed.), 9–21. Linnean Society of London, London. Reprinted from the *Biological Journal of the Linnean Society* 17(1), 1982.

Freeman, R. B. 1984. *Darwin Pedigrees*. Printed for the author, London.

Galton, Francis. 1869. *Heredity Genius: An Inquiry into its Laws and Consequences*. Julian Friedmann, London. Facsimile edition, 1978.

Ghiselin, Michael T. 2009. *Darwin: A Reader's Guide*. Occasional Papers of the California Academy of Sciences No. 155. California Academy of Sciences, San Francisco.

Glazebrook, R. T. 1928. [Obituary] Sir Horace Darwin, K.B.E., F.R.S. *Nature* 122: 580–581.

Glazebrook, R. T. 1928–1929. Sir Horace Darwin, 1851–1829. *Proceedings of the Royal Society of London, A* 122: vx–vxiii.
Glazebrook, R. T. 2004. Darwin, Sir Horace (1851–1928). *Oxford Dictionary of National Biography*, revised by Anita McConnell. Oxford University Press, Oxford.
Gluckman, Peter, Alan Beedle, and Marl Hanson. 2009. *Principles of Evolutionary Medicine*. Oxford University Press, Oxford.
Golubovsky M. 2008. Unexplained Infertility in Charles Darwin's Family: Genetic Aspect. *Human Reproduction* 23: 1237–1238.
Goodden, R. Y. 2004. Darwin, Sir Robert Vere [Robin] (1910–1974). *Oxford Dictionary of National Biography*. Oxford University Press, Oxford.
Gould, Stephen Jay. 1986. Knight Takes Bishop? *Natural History* 95: 18–33.
Hamilton, William D. 1972. Altruism and Related Phenomena, Mainly in the Social Insects. *Annual Review of Ecology and Systematics* 3: 193–232.
Hartmann, W. K. and D. R. Davis. 1975. Satellite-Sized Planetesimals and Lunar Origin. *Icarus* 24: 504–515.
Hayman, J. A. 2009a. Charles Darwin's Impressions of New Zealand and Australia, and Insights into His Illness and His Developing Ideas on Evolution. *Medical Journal of Australia* 191: 660–663.
Hayman, J. A. 2013. Charles Darwin's Mitochondria. *Genetics* 194: 1–5.
Healey, Edna. 2001. *Emma Darwin: The Inspirational Wife of a Genius*. Headline, London.
Healey, Edna. 2001. *Emma Darwin: The Inspirational Wife of a Genius*. Headline, London.
Henshilwood, Christopher S., et al. 2002. Emergence of Modern Human Behavior: Middle Stone Age Engravings from South Africa. *Science* 295: 1278–1280.
Huether, C. A., J. Ivanovich, B. S. Goodwin, E. L. Krivchenia, V. S. Hertzberg, L. D. Edmonds, D. S. May, and J. H. Priest. 1998. Maternal Age Specific Risk Rate Estimates for Down Syndrome among Live Births in White and Other Races from Ohio and Metropolitan Atlanta, 1970–1989. *Journal of Medical Genetics* 35: 482–490.
Huxley, Leonard (Ed.). 1900. *Life and Letters of Thomas Henry Huxley*. 2 vols. Macmillan, London.

Huxley, Thomas Henry. 1863. *Evidence as to Man's Place in Nature*. William & Norgate, London.

Huxley, Thomas Henry. 1882. [Obituary] Charles Darwin. *Nature*, 27 April. Reprinted in Huxley, Thomas Henry. 1893. *Darwiniana Essays*, 2: 244–247. Macmillan, London.

Huxley, Thomas Henry. 1888. [Obituary Notice:] Charles Robert Darwin. *Proceedings of the Royal Society of London* 44 (269): i–xxv. Reprinted in Huxley, Thomas Henry. 1893. *Darwiniana Essays*, 2: 253–302. Macmillan, London.

Jeffreys, Harold. 1924. *The Earth, Its Origin, History, and Physical Constitution*. Cambridge University Press, Cambridge.

Jensen, J. Vernon. 1988. Return to the Wilberforce-Huxley Debate. *British Journal for the History of Science* 21: 161–179.

Jones, John E., III. 2005. Memorandum Opinion. In the United States District Court for the Middle District of Pennsylvania. *Tammy Kitzmiller, et al. Plaintiffs v. Dover Area School Board Defendants*. 400 F. Supp. 2d 707, Docket No. 04cv2688.

Junker, Thomas. 2004. Darwin, Sir Francis (1848–1925). *Oxford Dictionary of National Biography*. Oxford University Press, Oxford.

Keith, Arthur. 1943. [Obituary] Major Leonard Darwin. *Nature* 151: 442.

Keynes, John Maynard. 1943. Leonard Darwin (1850–1943). *Economic Journal* 53: 438–439.

Keynes, Margaret E. 1943. Leonard Darwin (1850–1943). *Economic Journal* 53: 439–448.

Keynes, Randal. 2001. *Annie's Box: Charles Darwin, His Daughter, and Human Evolution*. Fourth Estate, London. US title: *Darwin, His Daughter & Human Evolution*.

Keynes, Randal. 2005. Darwin, William Erasmus (1839–1914). *Oxford Dictionary of National Biography*. Oxford University Press, Oxford.

Kuhn, Thomas S. 1962. *The Structure of Scientific Revolutions*. University of Chicago Press, Chicago.

Kuper, Adam. 2009. *Incest & Influence: The Private Life of Bourgeois England*. Harvard University Press, Cambridge, MA.

Kushner, David. 1993. Sir George Darwin and a British School of Geophysics. *Osiris* 8: 196–224.

Kushner, David. 2004. Darwin, Sir George Howard (1845–1912). *Oxford Dictionary of National Biography*. Oxford University Press, Oxford.

Lander, Eric S., et al. 2001. Initial Sequencing and Analysis of the Human Genome. *Nature* 409: 860–921.

Larmor, Joseph. 1912. Sir George Howard Darwin, K.C.B., F.R.S. *Nature* 90: 413–415.

Larson, Edward J. 1997. *Summer for the Gods: The Scopes Trial and America's Continuing Debate Over Science and Religion*. Basic Books, New York.

Larson, Edward J. 2004. *Evolution: The Remarkable History of a Scientific Theory*. Modern Library, New York.

Litchfield, Henrietta. 1910. *Richard Buckley Litchfield: A Memoir Written for His Friends*. Privately printed, Cambridge.

Litchfield, Henrietta. 1915. *Emma Darwin: A Century of Family Letters, 1792–1896*. 2 vols. John Murray, London. Kessinger reprint of D. Appleton edition, 2007.

Litchfield, Richard Buckley. 1903. *Tom Wedgwood, the First Photographer: An Account of His Life*. Duckworth, London.

Loy, James D. and Kent M. Loy. 2010. *Emma Darwin: A Victorian Life*. University Press of Florida, Gainesville.

Lucas, J. R. 1979. Wilberforce and Huxley: A Legendary Encounter. *Historical Journal* 22: 313–330.

Lyell, Charles. 1863. *Geological Evidences of the Antiquity of Man, with Remarks on Theories of the Origin of Species by Variation*. John Murray, London.

Lyons E. J., A. J. Frodsham, L. Zhang, A. V. S. Hill, and W. Amos. 2009a. Consanguinity and Susceptibility to Infectious Diseases in Humans. *Biological Letters* 5: 574–576.

Lyons E. J., et al. 2009b. Homozygosity and Risk of Childhood Death Due to Invasive Bacterial Disease. *BMC Medical Genetics* 10: 55, doi:10.1186/1471-2350-10-55.

Malthus, Thomas R. 1798. *An Essay on the Principle of Population, as It Affects Future Improvement of Society*. J. Johnson, London. [Darwin read the 1826 6th edition in 1838.]

Mawer, Simon. 2006. *Gregor Mendel: Planting the Seeds of Genetics*. Adams, New York.

McCalman, Iain. 2009. *Darwin's Armada: Four Voyages and the Battle for the Theory of Evolution*. W. W. Norton, New York.

Meacham, Standish. 1970. *Lord Bishop: The Life of Samuel Wilberforce, 1805–1873*. Harvard University Press, Cambridge, MA.

Mindell, D. P. 2009. Evolution in the Everyday World. *Scientific American* 300: 82–88.

Mindell, D. P. 2013. The Tree of Life: Metaphor, Model, and Heuristic Device. *Systematic Biology* 62: 479–489.

Mivart, St. George. 1871a. Darwin's *Descent of Man*. *Quarterly Review* 131: 47–90.

Mivart, St. George. 1871b. *On the Genesis of Species*. Macmillan, London.

[Mivart, St. George] Anonymous. 1874a. [Review of] *Researches into the Early History of Mankind and the Development of Civilization* by Edward Burnett Tylor, 1865; *Primitive Culture*, 1871; *Primitive Society*, 1873; *Prehistoric Times* by Sir John Lubbock, 1869; *The Origin of Civilisation and the Primitive Condition of Man*, 1870. *Quarterly Review* 137, No. 273: 40–77.

[Mivart, St. George] Anonymous. 1874b. [Response to George Darwin]. *Quarterly Review* 137, No. 274: 588–589.

Montgomerie, Robert. 2009. Charles Darwin's Fancy. *Auk* 126: 477–481.

Moody, J. W. T. 1971. The Reading of the Darwin and Wallace Papers: An Historical "Non Event." *Journal of the Society for Bibliography of Natural History* 5: 474–476.

Moore, James R. 1982. Charles Darwin Lies in Westminster Abbey. In *Charles Darwin: A Commemoration*. R. J. Berry (Ed.), 97–113. Linnean Society of London. Reprinted from the *Biological Journal of the Linnean Society* 17(1), 1982.

Moore, James R. 1989. Of Love and Death: Why Darwin "Gave Up Christianity." In *History, Humanity, and Evolution: Essays for John C. Greene*. James R. Moore (Ed.), 195–229. Cambridge University Press, Cambridge.

Moore, James. 1994. *The Darwin Legend*. Baker Books, Grand Rapids, MI.

Nature. 1925. [Obituary] Sir Francis Darwin, F.R.S. *Nature* 116: 583.

Nesse, Randolph M. and George C. Williams. 1996. *Why We Get Sick*. Vantage Books, New York.

Numbers, Ronald L. 2006. *The Creationists: From Scientific Creationism to Intelligent Design*. Expanded Edition. Harvard University Press, Cambridge, MA.

Ogilvie, Marilyn and Joy Harvey (Eds.). 2000. *The Biographical Dictionary of Women in Science: Pioneering Lives from Ancient Times to the Mid-20th Century*. Vol. 2. Routledge, New York.

Padian, Kevin. 2007. The Case of Creation. *Nature* 448: 253–254.

Paley, William. 1802. *Natural Theology; or, Evidences of the Existence and Attributes of the Deity, Collected from the Appearances of Nature*. Bridgewater Treatises, Faulder, London.

Pearson, K. 1914–1930. *The Life, Letters, and Labours of Francis Galton*, 3 vols. Cambridge University Press, Cambridge.

Pennock, Robert T. 2001. *Intelligent Design Creationism and Its Critics: Philosophical, Theological, and Scientific Perspectives*. MIT Press, Cambridge, MA.

Pietsch, Theodore W. 2012. *Trees of Life: A Visual History of Evolution*. Johns Hopkins University Press, Baltimore.

Pigliucci, Massimo. 2002. *Denying Evolution: Creationism, Scientism, and the Nature of Science*. Sinauer Associates, Sunderland, MA.

Porter, D[uncan] M. 1993. On the Road to the *Origin* with Darwin, Hooker, and Gray. *Journal of the History of Biology* 26: 1–38.

Porter, Duncan M. 2012. Why Did Wallace Write to Darwin? *Linnean* 28: 17–25.

Powell, Alan. 2009. *Far Country: A Short History of the Northern Territory*. Charles Darwin University Press, Darwin, NT, Australia.

Prodger, Phillip. 2009. *Darwin's Camera: Art and Photography in the Theory of Evolution*. Oxford University Press, Oxford.

Raverat, Gwen. 1952. *Period Piece: A Cambridge Childhood*. Faber & Faber, London.

Reeve, Tori. 2009. *Down House: The Home of Charles Darwin*. English Heritage, London.

Richmond, Marsha L. 2007. Box 2, (Emma) Nora Darwin Barlow (1885–1989), at the John Innes Horticultural Institute, 1913. *Nature Reviews, Genetics* 8: 897–902.

Ryde, Peter. 2004. Darwin, Bernard Richard Meirion (1876–1961). *Oxford Dictionary of National Biography*. Oxford University Press, Oxford.

Secord, J. A. 2000. *Victorian Sensation: The Extraordinary Publication, Reception, and Secret Authorship of "Vestiges of the Natural History of Creation."* University of Chicago Press, Chicago.

Sedgwick, Adam. 1845. Vestiges of the Natural History of Creation. *Edinburgh Review* 82: 1–85.

Shermer, Michael. 2002. *In Darwin's Shadow: The Life and Science of Alfred Russel Wallace*. Oxford University Press, Oxford.

Singer, Peter. 1977. *Animal Liberation: Towards an End to Man's Inhumanity to Animals*. Granada, London.

Sloan, Pat. 1960. The Myth of Darwin's Conversion. *Humanist* 75(3): 70–72.

Sloan, Pat. 1965. Demythologizing Darwin. *Humanist* 80(4): 106–110.

Spencer, Herbert. 1864. *Principles of Biology*. 2 vols. Williams and Norgate, London.

Spencer, Nick. 2009. *Darwin and God*. Society for Promoting Christian Knowledge, London.

Stearns, Stephen C. and Jacob C. Koella. 2008. *Evolution in Health and Disease*. 2nd edition. Oxford University Press, Oxford.

Stott, Rebecca. 2003. *Darwin and the Barnacle*. W. W. Norton, NY.

Stott, Rebecca. 2012. *Darwin's Ghosts: The Secret History of Evolution*. Spiegel and Grau, New York.

Symons, G. J. (Ed.). 1888. *The Eruption of Krakatoa and Subsequent Phenomena*. Report of the Krakatoa Committee of the Royal Society. Trübner, London.

Times [London]. 1925. [Obituary] Sir Francis Darwin. 21 September, 14.

Times [London]. 1928. [Obituary] Sir Horace Darwin. 24 September, 21.

Times [London]. 1943. [Obituary] Major L. Darwin. 27 March, 6.

Townshend, Emma. 2009. *Darwin's Dogs: How Darwin's Pets Helped Form a World-Changing Theory of Evolution*. Francis Lincoln, London.

Van Wyhe, John. 2009. *Darwin in Cambridge*. Christ's College, Cambridge.

Van Wyhe, John and Kees Rookmaaker. 2012. A New Theory to Explain the Receipt of Wallace's Ternate Essay by Darwin in 1858. *Biological Journal of the Linnean Society*. 105: 249–252.

Venter, J. Craig, et al. 2001. The Sequence of the Human Genome. *Science* 291: 1304–1351.

Voss, Julia. 2010. *Darwin's Pictures: Views of Evolutionary Theory, 1837–1874*. Yale University Press, New Haven, CT.

W., R. S. 1928. [Obituary] Sir Horace Darwin, K.B.E., F.R.S. *Cambridge Review*, October 26, 1928: 48–50.
Wallace, Alfred Russel. 1855. On the Law Which Has Regulated the Introduction of New Species. *Annals and Magazine of Natural History* 16: 184–196.
Wallace, Alfred Russel. 1905. *My Life: A Record of Events and Opinions*. 2 vols. Dodd, Mead, New York.
Wallace, Alfred Russel. 1908. *My Life: A Record of Events and Opinions*. 2nd edition, revised and abridged in 1 volume. Chapman and Hall, London.
Wedgwood, Barbara and Hensleigh Wedgwood. 1980. *The Wedgwood Circle: Four Generations of a Family and Their Friends*. Studio Vista, London.
White, Randall. 2003. *Prehistoric Art: The Symbolic Journey of Humankind*. H. N. Abrams, New York.
Wolpert, Lewis. 2007. *Six Impossible Things before Breakfast: The Evolutionary Origin of Belief*. W. W. Norton, New York.
Wright, Robert. 1994. *The Moral Animal*. Pantheon Books, New York.

ABOUT THE AUTHOR

Tim M. Berra is professor emeritus of evolution, ecology, and organismal biology at The Ohio State University, where he has been since 1972. He is also university professorial fellow at Charles Darwin University and research associate at the Museums and Art Galleries of the Northern Territory, both in Darwin, Northern Territory, Australia. He is a three-time Fulbright Fellow to Australia and has spent approximately 9 of the past 44 years doing field work on various species of fishes throughout Australia. He has taught at the University of Papua New Guinea and has been a visiting professor at the University of Concepción, Chile, and the University of Otago, New Zealand. He received a Ph.D. in biology from Tulane University, New Orleans, in 1969. He and his wife reside in a house of his own design with 16,000 books in the library, on a 20-acre property in Amish country near Bellville, Ohio, surrounded by a lake and a stream with 44 species of fishes.

Also by Tim M. Berra

William Beebe: An Annotated Bibliography (1977)
An Atlas of Distribution of the Freshwater Fish Families of the World (1981)
Evolution and the Myth of Creationism (1990)
A Natural History of Australia (1998)
Freshwater Fish Distribution (2001, 2007)
Charles Darwin: The Concise Story of an Extraordinary Man (2009)

INDEX

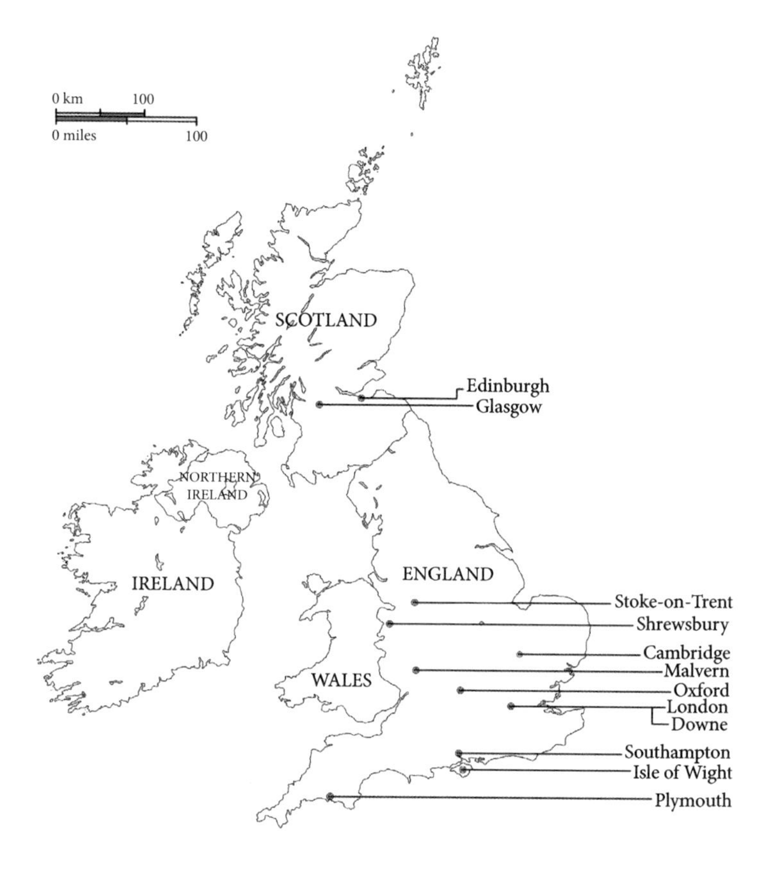
0 km
100
0 miles
100
SCOTLAND
Edinburgh
Glasgow
NORTHERN IRELAND
IRELAND
ENGLAND
WALES
Stoke-on-Trent
Shrewsbury
Cambridge
Malvern
Oxford
London
Downe
Southampton
Isle of Wight
Plymouth